SpringerBriefs in Electrical and Computer Engineering

More information about this series at http://www.springer.com/series/10059

Bo Rong · Xuesong Qiu
Michel Kadoch · Songlin Sun
Wenjing Li

5G Heterogeneous Networks

Self-organizing and Optimization

 Springer

Bo Rong
Communications Research Centre Canada
Ottawa, ON
Canada

Xuesong Qiu
Beijing University of Posts
  and Telecommunications
Beijing
China

Michel Kadoch
Department of Electrical Engineering
Ecole de Technologie Superieure
Montreal, QC
Canada

Songlin Sun
Beijing University of Posts
  and Telecommunications
Beijing
China

Wenjing Li
Beijing University of Posts
  and Telecommunications
Beijing
China

ISSN 2191-8112   ISSN 2191-8120   (electronic)
SpringerBriefs in Electrical and Computer Engineering
ISBN 978-3-319-39371-1   ISBN 978-3-319-39372-8   (eBook)
DOI 10.1007/978-3-319-39372-8

Library of Congress Control Number: 2016940361

Printed on acid-free paper

This Springer imprint is published by Springer Nature
The registered company is Springer International Publishing AG Switzerland

Preface

The wireless industry has already been preparing for the emerging fifth-generation (5G) standards in recent years. A 5G system is featured by (1) addressing the demands and business contexts of 2020 and beyond, (2) creating an access network to enable a fully mobile and interconnected society, (3) providing massive amount of miscellaneous devices with the connections to packet data network, and (4) bearing huge volume of data services in the manner of a deeply heterogeneous system. 5G achieves higher speed, increased capacity, decreased latency, and better quality of service (QoS). One of the promising technologies to meet the above requirement is the small cell network (SCN), which allows short-range, low-power, and low-cost base stations operating in conjunction with the main macrocellular network infrastructure. SCN introduces additional heterogeneity to wireless communication system, and thus, the so-called self-organizing network (SON) technology has been extensively studied to overcome the challenges.

5G network architecture tends to involve more low power nodes (LPNs), which make the system design more complicated. For example, the deployment of small cells in coverage holes can effectively reduce the penetration loss with a large amount of users. However, the small cell may produce interference to the users served by other power nodes. To solve this problem, we advocate a novel concept of smart LPN as the SON solution for 5G heterogeneous networks (HetNets). As illustrated in this book, the smart LPN integrates a number of hardcore technologies, such as cognitive radio and three-dimensional (3D) multiple-input and multiple-output (MIMO), to realize the function of automatic configuration and optimization. To provide physical layer insight into SON, 3D MIMO channel model has widely prevailed in recent years. 3D array deploys antenna elements in both horizontal and vertical dimensions. By taking into account the vertical dimension channel information, 3D beamforming not only provides more accurate channel estimation, but also mitigates intercell interference more effectively even without inter-eNB coordination.

The future 5G SON will integrate with a variety of cutting-edge technologies, for example, adopting the intelligent software-defined networking (SDN) and network function virtualization (NFV) to automatically take care of traffic control, resource allocation, density management, and security. SDN was formally defined by open networking forum (ONF), a user-driven organization dedicated to the promotion and adoption of SDN through open standards. SDN separates the control of network devices from the data they transport and the switching software from the actual network hardware. The OpenFlow standard integrates the network control plane into software running on an attached server or network controller, which enables the network control to become directly programmable and the underlying infrastructure to be abstracted for applications and network services. The NFV, on the other hand, is a network architecture that virtualizes entire classes of network node functions into building blocks. The goal of NFV is to decouple network functions from dedicated hardware devices and allow network services that are now being carried out by routers, firewalls, load balancers, and other dedicated hardware devices to be hosted on virtual machines (VMs). In this way, the operators can architect their networks by deploying network services on the top of standard devices. In the era of 5G, the network must be capable of meeting a huge amount of user diversified service demands at different data rates. Therefore, it should involve all the automatic technologies available, including SON, SDN, and NFV, to orchestrate all the available frequency resources, infrastructures, and hardware devices.

In 2010, Thomas L. Marzetta investigated multi-user MIMO (MU-MIMO) systems with multi-cellular time-division duplex (TDD) scenario, where each base station is supposed to deploy very large number of antennas, later known as massive MIMO. Massive MIMO has become one of the most important candidate technologies for 5G mobile networks. In massive MIMO system, the base stations (BSs) are equipped with hundreds of service antennas, in order to form diversified 3D MIMO channel between the BSs and users. With more antenna elements, it becomes realistic to steer the transmission toward the intended receiver and significantly reduce the interference. Massive MIMO can be deployed in heterogeneous networks, which are composed of a variety of cells, for example, a macrocell and several small cells. The cell association for users and the resource allocation for base stations become a problem due to complex network design and implementation. This raises the importance of huge spatial domain information management and complicated MIMO coordination.

The rest of this book is summarized as follows.

1. In Chap. 1, we highlight the important role of SON in 5G heterogeneous networks. First, several pivotal functions of SON are introduced and explained in the context of 5G scenarios. Then, cognitive radio, compressed sensing, and smart LPN are investigated as promising technologies for the implementation. Finally, numerical results are presented for the performance evaluation using 3D MIMO channel model.

2. In Chap. 2, we focus on the SDN and NFV technology for 5G heterogeneous networks. A variety of SDN standards are introduced and studied in details. An intelligent SDN architecture is then proposed based on the key technologies of 5G. To provide SDN with infrastructure support, NFV is also investigated as a flexible and affordable solution to service providers.

3. In Chap. 3, we investigate the 5G massive MIMO coordination under SDN platform. Massive MIMO technology is introduced as an essential part of future mobile networks. Then, an SDN-controlled MIMO coordination is proposed for the heterogeneous network environment. A null space-based hybrid precoding scheme is also developed for SDN implementation and evaluated with numerical results.

Contents

Acronyms

3D MIMO	Three-dimensional MIMO
3GPP	The Third Generation Partnership Project
5G	The fifth generation
ANR	Automatic neighbor relations
API	Application programming interface
BCS	Bayesian compressed sensing
BD	Block diagonal
BOSS	Business and Operation Supporting Systems
CAPEX	Capital expense
CCSA	China Communications Standards Association
CDMA	Code division multiple access
CPE	Control plane entity
CR	Cognitive radio
CS	Compressed sensing
CSI	Channel status indicator
DNS	Domain Name Service
DPC	Dirty paper coding
ETSI	European Telecommunications Standards Institute
FBMC	Filter-bank based on multi-carrier
FIR	Finite impulse response
GSM	Global System for Mobile Communications
HetNets	Heterogeneous networks
IETF	Internet Engineering Task Force
ITU-T	International Telecommunication Union Telecommunication Standardization Sector
LPN	Low power node
LTE	Long-term evolution
MIMO	Multiple-input and multiple-output
MIP	Mobile IP

MU-MIMO	Multi-user MIMO
NAT	Network address translation
NFV	Network function virtualization
ONF	Open networking forum
OPEX	Operating complexity and operating expense
QoE	Quality of experience
QoS	Quality of service
RAN	Radio access network
RATs	Radio access technologies
RF	Radio frequency
RIP	Restricted isometry property
RRM	Radio resource management
SBSs	Small cell base stations
SCN	Small cell network
SDN	Software-defined networking
SDR	Software-defined radio
SLA	Service level agreement
SON	Self-organizing network
SSL	Secure sockets layer
TDD	Time-division duplex
TD-SCDMA	Time-Division Synchronous Code Division Multiple Access
UE	User equipment
UPE	User plane entity
W-CDMA	Wideband Code Division Multiple Access
WiMAX	Worldwide Interoperability for Microwave Access
XaaS	Anything as a service

Chapter 1
Adaptive SON and Smart LPN for 5G Heterogeneous Networks

1.1 The Need of Self-organization in 5G

Just a few years after the first 4G smartphone hit the market, the wireless industry is already preparing for 5G. The main characteristics of 5G have been intensively studied recently, though the standards for 5G are not set yet. 5G will be faster, smarter and less power-hungry than 4G. It will introduce network virtualization, infrastructure sharing, concurrent operation at multiple frequency bands, and flexible spectrum allocations. With respect to network infrastructure, 5G has the features of network densification, miscellany of node types, simultaneous use of different medium access control and physical layers. Other than the move to 4G, which was all about boosting capacity, the move to 5G is about to create a network optimized for connecting billions of new devices to the Internet. As a result, the complexity and expense of operation can become the biggest challenge. Correspondingly, self-organizing has been investigated to cope with the challenges, e.g., data volumes are growing exponentially whereas the average revenue per user has remained more or less flat.

Small Cell Network (SCN) is a promising technology in 5G network, based on the idea of deploying short-range, low-power, and low-cost base stations operating in conjunction with the main macrocellular network infrastructure. Market projections suggest a rapid increase in the number of small cells deployed over the coming years, as it can enable next-generation networks to provide high data rates by offloading traffic from the macro cell. Many operators plan to set up 3–5 micro- or pico-BSs per macrocell to meet the huge amount of data capacity requirements. Peak numbers even state 13–20 picos per macrocell, at least in urban environments [1].

Different from macrocells, the amount of traffic captured by small cells is not strongly related to the size of their coverage area. The optimal number of small cells to be deployed varies with the fraction of traffic actually served, instead of considering merely the number of small cells deployed per macrocell. According to Cisco's Global Mobile Traffic Forecast [2], the fraction of overall mobile traffic

© The Author(s) 2016
B. Rong et al., *5G Heterogeneous Networks*,
SpringerBriefs in Electrical and Computer Engineering,
DOI 10.1007/978-3-319-39372-8_1

offloaded to both femtocells and WiFi access points was about 30 % in 2012 and will grow up to 46 % globally until 2017. Other data source suggests that in certain European and North American networks already 50–60 % of today's wireless traffic is being offloaded to WiFi access points [3]. Predictions by Juniper research indicate that up to 60 % mobile data traffic may be offloaded to microcells and picocells as well as to femtocells and WiFi as early as 2016 [4].

Small cells are available for a wide range of air interfaces, including GSM, CDMA2000, TD-SCDMA, W-CDMA, LTE and WiMax, Small cells are equivalent to macro-BSs apart from the transmit power. However, there are two key differences between small cells and conventional macro cells considering the network architecture.

The first difference is that we need much more of small cells everywhere than macro cells—indoors outdoors, at homes, offices, enterprises, shopping malls and lamp posts. In consequence, manual processes for configuration and optimization are no longer feasible. We have to bring the benefits of more ad hoc deployments to all scenarios, in order to cost-effectively deploy small cell on a large scale. Dense deployments can pose multiple configuration, mobility and coordination challenges. Usually configuration and optimization begins with acquisition of an IP address, downloading the correct software version, and then downloading the correct parameter configuration from an operator's data base. The process continues with adjusting parameter configuration during operation based on measurements, key performance indicators (KPIs), customer complaints, etc. All those aspects must be self-organized to realize the vision of plug-and-play, self-configuring and self-optimizing for small cell networks.

The second difference is that small cells are deployed much more dynamically. Whereas macrocell deployments are thoroughly planned, it is important to adopt a more 'unplanned' ad hoc deployment approach to rapidly achieve hyper-densification. Small cells are often commissioned quickly whenever capacity need is detected and there is no need for detailed radio frequency (RF) network planning and optimization. Since, in most cases, small cells do not provide basic network coverage, operators may tend to switch them off when capacity is not needed. These small cells could be operator or user installed, but managed by operators, and coordinate with macros as well as other small cells. In general, the impact of small-cell deployments on the users connected to macrocells must be kept minimal all the time.

Small cells have to be plug-and-play and self-configurable. This requirement highlights the importance of self-organizing network (SON) for SCNs. SON was originally codified within 3GPP Release 8 and proven to be very valuable already for traditional macro networks, to adapt, e.g., to changes in the environment, the traffic conditions, or the deployment [5]. In 5G, SON will continue playing an essential part, as heterogeneous networks (HetNets) require even more automatic control.

1.2 SON for 5G Mobile Networks

Though the concept of HetNets has been a huge success, it is becoming increasingly clear that if 5G is to become a viable reality, network operators are going to adopt a variety of technologies. Consequently the operating complexity and operating expense (OPEX) will be the biggest challenge due to such a conglomeration of technologies.

To overcome this difficulty, one logical approach is to deploy SON in 5G network to achieve fully self-organizing with end-to-end network behavior intelligence. The network can then exploit the cognition of the network state to divert and focus the right amount of network resources when and where needed. This way, the users are able to always perceive seamless and limitless connectivity. The most basic concept of self-organizing networks allows base stations to automatically configure themselves, i.e., automatic neighbor relations (ANR), which will be much further developed in 5G systems. Major capacity gain of 5G has to come from mostly impromptu densification and network-level efficiency enhancement [6], and thus the technical viability of future wireless networks almost exclusively hinges on the evolution of SON paradigm.

In SON, there are three main functions, which replace the classic manual configuration, post deployment optimization, and maintenance in traditional networks respectively.

(1) Self configuration: The aim is for base stations to become essentially "Plug and Play" items. They should need as little manual intervention in the configuration process as possible. This will enable the skill level of installers to be reduced, thereby saving costs while improving the reliability. Accordingly this is a major element within the overall self-organizing network, SON software.

(2) Self optimization: Once the system has been set up, it will be necessary to optimize the operational characteristics to best meet the needs of the overall network. This is achieved by self-optimization routines within the overall self-organizing network, SON software.

(3) Self-healing: Any system will develop faults from time to time. This can cause major inconvenience to users. However, it is often possible for the overall network to change its characteristics to temporarily mask the effects of the fault. Boundaries of adjacent cells can be increased by increasing power levels and changing antenna elevations, etc. This self-healing aspect of SON is of great interest.

When operating concurrently in the same network, different SON functionalities can have parametric or objective-based conflicts, which may undermine the overall gains of SON. Therefore, the Third Generation Partnership Project (3GPP) has emphasized the self-coordination among SON functions to ensure stable network operation. However, it remains an under-addressed problem even for 3G and 4G. As 5G network architecture tends to become more complex, the analysis of potential conflicts generated by the numerous autonomous SON functionalities and the design of an appropriate self-coordination framework can be extremely challenging. Therefore, 5G self-coordination has to be arranged from the grassroots level of the SON functions' design. And this is very different from 4G, where a retrospective approach has been taken to embed self-coordination into relatively independently developed SON functionalities. The first step toward designing conflict-free SON functions for 5G may start from a comprehensive identification and taxonomy of potential conflicts in SON [7].

Moreover, since the 5G targets at creating perception of zero latency, the 3G/4G type of reactive SON will not be able to meet the performance requirements of 5G. This is due to the fact that, in classic SON, it causes latency to observe the situation, diagnose the problem, and then trigger the compensating action, which cannot meet with 5G targeted quality of experience (QoE) levels. Therefore, to be able to perform successfully in 5G network, the intended solution for SON paradigm should be proactive. In this way, the network can predict the potential problem beforehand instead of waiting to observe and spot the problem. This proactive SON can be achieved by inferring network-level intelligence from the massive amount of control, signaling, and contextual data that can be harnessed in mobile networks to predict the problem in its infancy, and then take preemptive actions to resolve the problem before it occurs. This approach can substantially reduce the intrinsic delay between the observation and compensation phases compared to current state-of-the-art SON even if not all problems can be predicted beforehand. Empowering SON with big data is the key to transforming SON from being reactive to proactive.

In the following, we will propose a novel concept of smart low power node (LPN) as the SON solution for 5G HetNets. As illustrated later, the smart LPN integrates a number of hardcore technologies, such as cognitive radio and 3D multiple-input and multiple-output (MIMO), to realize the function of automatic configuration and optimization.

1.3 Cognitive Radio and Compressed Sensing

Cognitive radio (CR) is a form of intelligent communication system in which a transceiver can intelligently detect which communication channels are in use and which are not, and instantly move into vacant channels while avoiding occupied ones. CR has the potential of detecting the variation of wireless environment nearby and making the corresponding action responding to the change, which optimizes the use of available radio-frequency (RF) spectrum while minimizing interference to other users. Although cognitive radio can sense very broad frequency band, there's a challenge in wideband sensing because of the high RF signal acquisition costs. Most of the existing spectrum detection algorithms are based on per-channel sensing and thus require very high sampling rate that lowers the system efficiency and makes detecting the spectrum holes promptly very difficult. To address this problem, the emerging compressed sensing (CS) theory has been applied in cognitive radio which can provide reliable spectrum sensing at affordable complexity.

Compressed sensing takes less samples than Nyquist sampling and is an innovative frame for wideband spectrum sensing in CR field. Previous work in [8] has developed compressed sensing techniques tailored for the coarse sensing in spectrum holes detection, which uses Sub-Nyquist rate samples to sense and classify bandwidths by a wavelet based edge detector. A Bayesian compressed sensing (BCS) framework has been utilized in [9] to sample sparse signals at sub-Nyquist rates in CR system for alleviating the bandwidth constraints imposed by frontend analog-to-digital converters (ADCs). With capturing the signal at sub-Nyquist rates, compressed sensing makes it possible to reconstruct a sparse signal and thus wideband spectrum sensing is doable by compressed sensing (CS).

In CS, a signal with a sparse representation can be recovered from a small set of considerably fewer linear measurements, which can be considered of estimating a sparse vector $x \in C^N$ from an observed vector of measurements $y \in C^M$ based on the linear model ("measurement equation"). We can represent this process mathematically as:

$$\mathbf{y} = \Phi \mathbf{x} + \mathbf{w} \tag{1.1}$$

where $\Phi \in C^{M \times N}$ is a known measurement matrix and $\mathbf{w} \in C^N$ is an unknown vector that represents measurement noise and modeling errors. Typically, the reconstruction process can be operated ideally only when signal $\mathbf{x}$ is S-sparse ($S \ll N$), which implies that there can be at most S nonzero entries. We should design the matrices Φ with $N \gg M$ to ensure that we will be able to recover the original signal accurately and efficiently. And actually the number of variables to be estimated is much larger than the number of observations, which satisfies this condition.

However, the signal to be recovered is usually not sparse in practical applications, but can be represented in some basis $\{\Psi\}_{i=1}^N$ with the corresponding sparse coefficients θ_i. Then Eq. (1.1) can be expressed as

$$\mathbf{y} = \Phi \mathbf{x} + \mathbf{w} = \Phi \Psi \theta + \mathbf{w} = \Theta \theta + \mathbf{w} \tag{1.2}$$

where $\theta \in C^L$ is a L dimension sparse vector of coefficients based on an basis matrix $\Psi \in C^{N \times L}$, and $\Theta = \Phi \Psi$ is a $M \times L$ matrix ($M \ll L$)

To enforce the sparsity constraint of solving the underdetermined linear equations and avoid solution's indeterminacy that arises from much fewer measurements compared with entries in θ, the S-sparse signal to be recovered, l_0 norm minimization which aims to minimize the number of nonzero components of the solution can be utilized for the reconstruction process as follows.

$$\hat{\theta} = \min \|\theta\|_0, \quad s.t. \|\Theta \theta - y\|_2 \leq \Xi \tag{1.3}$$

where $\hat{\theta}$ refers to the estimated vector for θ and $\|\mathbf{w}\| \leq \Xi$ is noise tolerance. However, the l_0 norm minimization is a NP-hard problem and cannot satisfy the requirements of the practical applications. As a solution, a variety of low-cost sparse recovery algorithms are being developed.

Obviously, the reconstruction of a compressible signal x can be reliable only when matrix Θ cannot map two different S-sparse signals to the same set of samples, implying the matrix Θ should be designed to satisfy the Restricted Isometry Property (RIP) [10].

1.4 3D Channel Model for 5G HetNets

Recently, three-dimensional MIMO (3D MIMO) has been highly-focused as a promising technique to enhance vertical coverage and the entire system performance, which deploys antenna elements in both horizontal and vertical dimensions. The 3D MIMO channel model is presented in [11], where the vertical dimension channel information is taken into account. 3D beamforming techniques can mitigate inter-cell interference more effectively even without inter-eNB coordination. It is of practical interest to assess the capacity of 3D MIMO system. The employment of higher order codebook can promote the system capacity for a fixed number of antennas per cell site [12], and under such condition, the inter-cell interference becomes more spatially at a less performance loss.

In this book, we propose a cognitive MU-MIMO scheduling scheme by taking into account of 3D MIMO channel, compressed sensing based inter-cell interference coordination and 3D codebook optimization. We adopt an angle based compressed sensing algorithm to acquire the spectrum usage information for 3D codebook design, resource allocation and user scheduling. By applying CS in cognitive CR for spectrum sensing, our proposed scheme can effectively acquire the use of the frequency band, and then perform an adaptive frequency angle selection algorithm for scheduling multiple users with optimized 3D codebook. The numerical simulation results demonstrate that the proposed scheme can offer a complexity reduction and considerably improve the system performance.

Our proposed 3D MU-MIMO scheduling scheme has twofold of innovation. Firstly, it employs 3D angle based compressed sensing. Secondly, it performs the resource allocation and user scheduling based on 3D codebook. In HetNets, small cells adopt cylindrical antenna to acquire the knowledge of interferences, user distribution and spectrum usage. The employment of high order codebooks can significantly simplify the user selection procedure to reduce computational complexity while maintaining sufficient system capacity.

The channels are more completely described with the aid of 3D channel considering both the azimuth and elevation angle information of BSs. The elevation angle is also involved in the channel models standardization and draws much attention. Moreover, the joint and separate 3D codebook subset is designed to perform horizontal and vertical beamforming by utilizing the channel correlation in spatial domain.

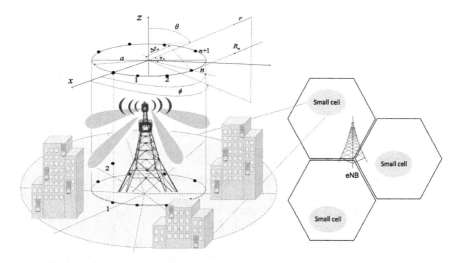

Fig. 1.1 HetNets with the application of cylindrical array in small cell

Here we consider a multi-cell HetNet where one eNB and several small cells coexist in the same coverage area and share channel frequency bands [13]. As shown in Fig. 1.1, we equally divide the small cell coverage into N sections corresponding to the antennas distribution. This way, the rank deficiency problem can be avoided.

Particularly, we assume a basic model of uniform cylindrical array, where a small cell is equipped with $M \times N$ antenna elements serving K single-antenna users. Different from linear and circular arrays, cylindrical array provides a 3D angular scan, in both horizontal φ and vertical θ. The cylindrical array consists of M tiers of circular arrays and each array has N antenna elements. From vertical perspective, it can also be considered as N uniform linear arrays with M antenna elements for each. The element spacing is d. From a geometric point of view, it is clear that an outgoing wave at the mth element leads the phase at the $(m+1)$th element by $kd\sin\theta$, where $k = 2\pi/\lambda$. Then the array factor can be obtained as:

$$AF_V = 1 + e^{j(kd\cos\theta + \beta)} + e^{j2(kd\cos\theta + \beta)} + \cdots + e^{j(M-1)(kd\cos\theta + \beta)} \qquad (1.4)$$

From the horizontal perspective, we consider a uniform circular array with N elements and a current phase of β_n (reference to the central point of the array) for the nth element (and $\phi_n = 2\pi n/N$). Then we can obtain the array factor as:

$$AF_H = e^{j[ka\sin\theta\cos(\phi-\phi_1) + \beta_1]} + e^{j[ka\sin\theta\cos(\phi-\phi_2) + \beta_2]} + \cdots + e^{j[ka\sin\theta\cos(\phi-\phi_N) + \beta_N]}$$

$$= \sum_{n=1}^{N} e^{j[ka\sin\theta\cos(\phi-\phi_n) + \beta_n]}$$

$$(1.5)$$

Therefore the joint array factor of cylindrical array is obtained by

$$AF = AF_H \otimes AF_V \tag{1.6}$$

where $\otimes$ denotes the Kronecker product. The array factor of cylindrical array can be calculated as

$$AF = \sum_{m=0}^{M-1}\sum_{n=0}^{N-1} e^{-j\frac{2\pi}{\lambda}(a\cos(\phi_n-\beta_H)-a\cos(\phi_n-\phi_0)+md\sin\beta_V-md\sin\theta_0)} \tag{1.7}$$

Figure 1.2 illustrates our proposed scheme where the small cell is equipped with a cylindrical array. The small cell acquires the spectrum information by utilizing the angle based compressed sensing algorithm and then detects the interference from macro BS and small cells nearby. With the channel information detected, the proposed scheme performs resource allocation for users with an adaptive spectrum selection algorithm. Furthermore, the specially designed 3D common base codebook can store the information of the devices and improve the system efficiency.

In the proposed system, the small cell equipped with M × N transmit antennas needs to be actuated by K radio frequency chains. It is worth noting that KRF is supposed to be far smaller than M × N, which restricts the maximum number of transmitted streams. We assume that there are exactly KRF single-antenna users to be scheduled and each user supports a single-stream transmission. Hence, the downlink precoding becomes RF processing. The RF precoder is denoted by F of dimension M × N × KRF. We assume that the RF precoder performs only phase control with variable phase shifters and combiners. Each entry of F is normalized to satisfy $|F_{i,j}| = \frac{1}{\sqrt{M \times N}}$, where $F_{i,j}$ denotes the (i, j)th element of F.

We adopt the narrowband flat fading channel and the sampled baseband signal received at the kth user is obtained by

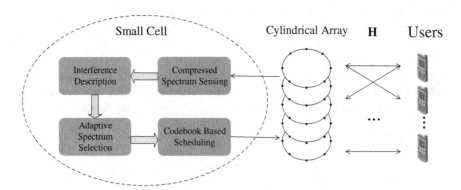

Fig. 1.2 Proposed cognitive 3D MU-MIMO scheduling scheme in cylindrical array

$$y_k = h_k^H Fs + n_k \tag{1.8}$$

where h_k^H denotes the downlink channel between the small cell and the kth user. And $s \in Ck \times 1$ is the signal vector for a total of K users which satisfies $E[ss^H] = \frac{P}{k}I_K$, where P is the transmit power at the small cell and E $[\cdot]$ is the expectation operator. n_k is the additive noise, which is assumed to be circular symmetric Gaussian with unit variance, i.e., $n_k \sim CN (0, 1)$.

The key procedure to perform the phase-only control in RF domain is extracting phases from the conjugate transpose of the aggregate downlink channel between the small cell and multiple users. In our proposed algorithm, the phases of the channel elements are adjusted to achieve the largest array gain provided by excessive antennas. Moreover, 3D precoding takes account of one more dimension than traditional 2D precoding, which only distinguishes users of different horizontal dimension. It means that our proposed 3D precoding scheme can distinguish users not only from horizontal but also from vertical plane. Hence, we can distinguish users of the same horizontal dimension from the vertical dimension. In this way, the beam for each user will be more precisely targeted and the energy will be more focused, which consequently reduce the interference to the other users. To clarify, we begin with the 2D RF precoding and then expand it to 3D dimension.

Firstly, we calculate the precoding codeword from horizontal direction:

$$F_{i,j}^H = \frac{1}{\sqrt{N}}e_k^{j\beta_{i,j}^H} \tag{1.9}$$

where $\beta_{i,j}$ is the phase of the (i, j)th entry of $[h_1,\ldots,h_k]$, which is the conjugate transpose of the horizontal composite downlink channel.

Similarly, the vertical plane 2D codeword in the codebook can be written as

$$F_{i,j}^V = \frac{1}{\sqrt{M}}e_k^{j\beta_{i,j}^V} \tag{1.10}$$

To achieve the joint 3D codebook, we need to employ Kronecker product of vertical and horizontal codeword. Based on Eqs. (1.9) and (1.10), the Kronecker product 3D codebook F can be written as

$$F = F^V \otimes F^H \tag{1.11}$$

where $\otimes$ denotes the Kronecker product and $F = \{f_0, f_1,\ldots,f_{m,n},\ldots,f_{M\times N-1}\}$. f_m^V and f_m^H represent the mth and nth codeword in vertical and horizontal codebook respectively.

Since the channels are highly spatially correlated as mentioned above, each element of the codebook corresponds to a combined beamformed angle. With the codebook size increasing, the corresponding angles will be more and finer. Meanwhile, the need to feedback more bits with lager codebook leads to more system overhead. Hence we should choose appropriate codebook size considering the tradeoff between beamforming accuracy and feedback overhead.

1.5 Smart LPN for 5G HetNets

The deployment of small cells in coverage holes can effectively reduce the penetration loss with a large amount of users. However, the small cell may produce interference to the users served by other power nodes, and simultaneously be affected by the interference from surrounding small cells and macro cells. To solve this problem, a synthetic interference matrix is utilized in resource allocation in our proposed algorithm, which is generated by the frequency and power distribution in different directions collected by the small cell. The synthetic interference matrix is extended by the spectrum sensing matrix in the system.

Figure 1.3 describes a basic model, where an integrated cell consisting of three sectors is surrounded by several power nodes. We can get the position information of all the power nodes from operator's database. In this procedure, the interferences from other eNBs and small cells are taken into account. We can fulfill the modelling

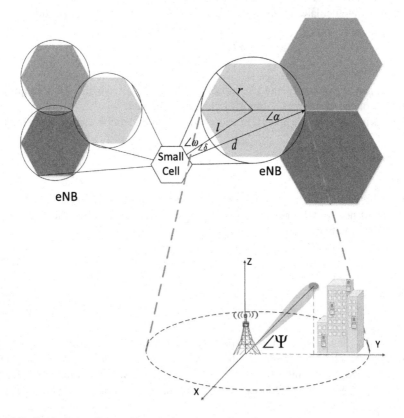

Fig. 1.3 Synthetic interference by a single power node

process by figuring out the key parameters, such as cell radius r, the distance d between small cell and the eNB, and the angle α. Reference [14] provides a reliable sensing algorithm, which is preferable to be referenced in our scheme with cylindrical antenna and 3D MIMO techniques applied. In the following text, we will illustrate the steps of computing the extensional angle.

First we have the geometric relationship

$$l = \sqrt{r^2 + d^2 - 2rd\cos\alpha}. \tag{1.12}$$

Then the angle δ can be calculated as

$$\delta = \arccos\left(\frac{d^2 + l^2 - r^2}{2dl}\right) \tag{1.13}$$

Likewise, the angle ω can be obtained by

$$\omega = \arcsin\left(\frac{r}{l}\right) \tag{1.14}$$

We define the extension angle ω_1 and ω_2 as

$$\omega_1 = \omega + \delta, \quad \omega_2 = \omega - \delta. \tag{1.15}$$

Finally, to be consistent with [15], let θ be the horizontal angle and φ be the vertical angle. We expand the scheme to 3D dimension by considering the vertical extension angle ψ as

$$\Psi = \arctan\frac{z}{d} \tag{1.16}$$

Then we can expand the spectrum at angle θ with ω_2, ω_1 and ω as shown in Fig. 1.4. For instance, we can keep the elements at the θ_ith row unchanged and add the elements from the $(\theta_i - \omega_1)$th row to the $(\theta_i + \omega_2)$th row by the elements at the θ_ith row. Similarly, the interference to any general small cells can be achieved in the same way by Eqs. (1.12)–(1.16).

We define a frequency-angle sensing matrix $A_{i \times j \times k}$ as the synthetic interference matrix to avoid multi-inference. Here i and j represents the index of angles in horizontal and vertical divisions respectively, and k represents the index of frequency bands. The matrix A takes account of all the G power nodes interference matrix $B_g(g = 0, 1, 2,...,G - 1)$ and is given by the superposition of all B_g

$$A = \sum_{g=0}^{G-1} B_g \tag{1.17}$$

The elements of A are empty, when there are no power nodes at the corresponding angles or frequency bands. Using matrix A, one can easily develop an

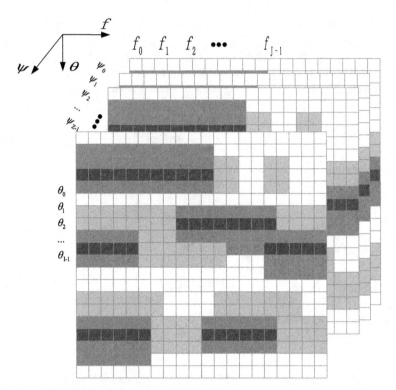

f : Frequency band

ψ : Vertical extension angle

θ : Horizontal extension angle

Fig. 1.4 Synthetic interference matrix A

adaptive interference detection and frequency angle selection scheme, which can optimally choose the frequency to use for a small cell. A detailed algorithm can be found in [15].

Once finishing the frequency allocation, the next task is to consider the multiuser MIMO scheduling for the small cell. The small cells and users share the same codebook to reduce system overhead. We can obtain the codebook by Grassmannian subspace packing to maximize the minimum distance among subspaces spanned by codebook matrices. In the meantime, the users adjust to match the codebook elements and group themselves according to the channel status. Within one user group, the scheduling process is based on relative channel quality, i.e., the instantaneous channel quality condition of the subscriber divided by its current average throughput. Within one time slot, the subscriber with the largest relative channel quality is to be selected. The multiuser MIMO system can achieve throughput-fairness tradeoff by considering the relative channel quality and codebook.

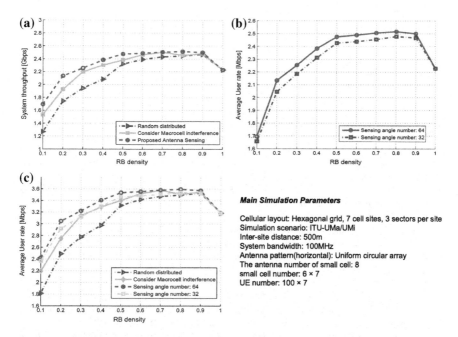

Fig. 1.5 Performance evaluation. **a** System throughput using different schemes. **b** Average user data rate with different numbers of sensing angles. **c** Average user data rate with different schemes

To evaluate the performance of the proposed scheme above, we simulate three schemes the small cells can be deployed in contrast: the scheme with randomly distributed RBs, the scheme with RBs allocated to each user concerning the interference from macro cells, and our proposed scheme using the compressed sensing based on cylindrical antenna array.

Figure 1.5a shows that our proposed scheme can significantly enhance the overall system throughput. Furthermore, the throughputs of all three schemes tend to be similar when the density of RBs increases. This is due to the fact that larger density of RBs causes an increased probability of the inevitable interference. It also demonstrates that the increase of RBs density enhances the system throughput. Figure 1.5b reveals that the system performance is enhanced along with the increase of sensing angles numbers in the proposed system. The system can deploy more subscribers at a single time slot when sensing with more angles. Figure 1.5c compares the average user data rate with different schemes and different sensing angle numbers. We can see the results can satisfyingly match our expectations.

References

1. R. Webb, T. Wehmeier, K. Dyer, Small Cells 2012 Integration and Optimisation. Mobile Europe, Technical Report (2012)
2. Cisco, Cisco Visual Networking Index: Global Mobile Data Traffic Forecast Update 2012–2017. Cisco Inc., Technical Report (2012)
3. RadioOpt GmbH, RadioOpt (2012)
4. N. Bhas, Mobile Data Offload & Onload. Juniper Research, Technical Report (April 2012)
5. S. Hamalainen, H. Sanneck, C. Sartori (eds.), *LTE Self-Organizing Networks* (Wiley, New York, 2012)
6. N. Bhushan et al., Network densification: the dominant theme for wireless evolution into 5G. IEEE Commun. Mag. **52**(2), 82–89 (2014)
7. H.Y. Lateef, A. Imran, A. Abu-dayya, in *A Framework for Classification of Self-Organising Network Conflicts and Coordination Algorithms.* Proceedings of IEEE PIMRC'13 (2013), pp. 2913–2918
8. Z. Tian, G. Giannakis, in *Compressed Sensing for Wideband Cognitive Radios.* International Conference on Acoustics, Speech and Signal Processing, pp. lV/1357-IV/1360, Apr 2007
9. S. Hong, in *Direct Spectrum Sensing from Compressed Measurements.* IEEE MILCOM, pp. 284–289, Oct 2010
10. E.J. Candes, The restricted isometry property and its implications for compressed sensing. C. R. Math. Acad. Sci., Serie I **346**, 589–592 (2008)
11. F. Peng, Y. Wang, W.D. Zhang, 3D-MIMO codebook aided multiple user pairing scheme in LTE Advanced systems. J. China Univ. Posts Telecommun. **21**, 1–9 (2014)
12. Y. Yuan, Y. Wang, W. Zhang, F. Peng, in *Separate Horizontal Amp; Vertical Codebook Based 3D MIMO Beamforming Scheme in LTE-A Networks.* Vehicular Technology Conference, pp. 1–5, Sept 2013
13. R.Q. Hu, Y. Qian, *Heterogeneous Cellular Networks* (Wiley, London, 2013)
14. S. Sun, M. Kaoch, T. Ran, in *Adaptive SON and Cognitive Smart LPN for 5G Heterogeneous Networks.* Mobile Networks and Applications: Special Issue on Networking 5G Mobile Communications Systems: Key Technologies and Challenges, Jan 2015
15. F. Qi, S. Sun, B. Rong, R.Q. Hu, Y. Qian, *Cognitive Radio Based Adaptive SON for LTE-A Heterogeneous Networks* (IEEE GLOBECOM, 2014), pp. 4412–4417

Chapter 2
Intelligent SDN and NFV for 5G HetNet Dynamics

2.1 Envision of 5G Mobile Networks

A technology breakthrough happens in mobile communications almost every ten years. 5G, as an emerging example right now, provides not only simply faster speed but also increased capacity, decreased latency and better quality of service (QoS). To meet the 5G system requirements, it needs a dramatic change in the design of cellular architecture. Next we will present a detailed survey on the 5G cellular network architecture, and address some of the key emerging technologies helpful to improve the efficiency and meet the user demands.

Although there has been a significant debate on what 5G exactly is, most researchers believe it is not necessary to have a giant change in the wireless setup like what had happened from 1G to 4G. For example, current technologies like OFDMA will work at least for next 50 years. Alternatively, service providers can implement the advance technology at the fundamental network to adopt the value-added services to please user requirements. This will provoke the package providers to drift for a 5G network as early as 4G is commercially set up [1].

5G network elements and terminals are highly advanced and characteristically upgraded to afford future scenarios. A general demographic in [2] indicates that most wireless users stay indoors for approximately 80 % of time and stay outdoors for approximately 20% of time. With current wireless cellular architecture, an outdoor base station in the middle of a cell helps mobile users communicate both indoor and outdoor. For indoor users to communicate with the outdoor base station, the signals will have to travel through the walls of the indoors, resulting in high penetration loss, reduced data rate, low spectral efficiency. To overcome this challenge, one of the key ideas for scheming the 5G cellular architecture is to separate outdoor and indoor scenarios. This way, the penetration loss through building walls can significantly be avoided. The idea will be assisted by the massive MIMO technology [3], in which geographically dispersed array of tens or hundreds of antennas are deployed. Different from the conventional MIMO systems using

© The Author(s) 2016
B. Rong et al., *5G Heterogeneous Networks*,
SpringerBriefs in Electrical and Computer Engineering,
DOI 10.1007/978-3-319-39372-8_2

either two or four antennas, massive MIMO system has come up with the idea of utilizing the advantages of large array antenna elements in terms of huge capacity gains.

We have the following steps to build up a large massive MIMO network. First, the outdoor base stations will equip with large antenna arrays, and among them, some are dispersed around the hexagonal cell and linked to the base station through optical fiber cables. Second, large antenna arrays will also be installed on top of each building to communicate with outdoor base stations in line of sight. Large antenna arrays on top of building then attach to the wireless access points inside the building through cables, in order to improve the energy efficiency, cell average throughput, data rate, and spectral efficiency. In this infrastructure, the indoor users will only have to connect or communicate with indoor wireless access points [1]. For indoor communication, many technologies can be utilized for short-range communications with high data rates, including WiFi, ultra wideband, millimeter wave communications [4], and visible light communications [5]. It is noteworthy that technologies, such as millimeter wave and visible light communication, utilize higher frequencies and are not conventionally used for cellular communications. It is usually not efficient to use these high frequency waves for outdoor and long distance applications, because these waves will not infiltrate from dense materials efficiently and can be easily dispersed by rain droplets, gases, and flora. Along with the introduction of new spectrum, there exists one more method to solve the spectrum shortage problem, i.e., cognitive radio, which changes frequencies, modulation schemes and other parameters under the control of spectrum databases, radio frequency (RF) sensing technology or both [6].

In 5G heterogeneous networks, the architecture includes relays and different sizes of cells referred to as macro-, micro-, pico- and femto-cells. Current research efforts have considered deploying mobile small cell networks as an integral entity that partially comprises of mobile relay and small cell concepts [7]. Mobile small cells support the users in high mobile networks, such as automobiles and high speed trains, to enhance vehicular users' QoS. A mobile small cell has the antennas located inside vehicle to communicate with the onboard users. In the meantime, it also has the massive MIMO unit consisting of large antenna arrays placed on the top of vehicle to communicate with the outside base station. From users' perspective, a mobile small cell is seen as a regular base station, which helps receive stronger signal and obtain higher data rate with considerably reduced signaling overhead, as shown in [1, 7].

5G network architecture consists of two logical layers: a radio network and a network cloud. The radio network consists of different types of components performing different functions. The network function virtualization (NFV) cloud consists of a user plane entity (UPE) and a control plane entity (CPE) performing higher layer functionalities related to the user and control plane respectively. It helps service providers accelerate the rollout of new services and reduce capital expense (CAPEX)/operating expense (OPEX). Anything as a service (XaaS) will provide service as per need, with resource pooling as one of the examples [8].

The proposed 5G network architecture in [1, 8] highlights the interconnectivity among the different emerging technologies such as massive MIMO network, cognitive radio network, mobile and static small-cell networks. As stated, front end and backhaul network have equal importance in 5G network architecture. This architecture also explains the role of network function virtualization (NFV) cloud in the 5G mobile network. In general, the proposed architecture may provide a good platform for future 5G standardization network.

As discussed above, the future 5G system will integrate a variety of cutting edges technologies. Correspondingly, we will propose in the following a novel concept of intelligent Software-Defined Networking (SDN) and NFV to automatically take care of traffic control, resource allocation, density management, security, etc.

2.2   SDN Technology Overview

SDN is an emerging architecture which was first introduced in the 1990s and became popular in the 21st century. The architecture of SDN was formally defined by open networking forum (ONF) which is a user-driven organization dedicated to the promotion and adoption of SDN through open standards development [9]. ONF is also responsible for the maintenance of the OpenFlow standard and technical specifications for the OpenFlow switch as well as the conformance test of SDN enabled devices. SDN separates the control of network devices from the data they transport, and the switching software from the actual network hardware. It has three layers: application, control and infrastructure. The functions of two lower layers are called OpenFlow controller and OpenFlow switch, corresponding to the control and data planes of traditional IP/MPLS network switches and routers. OpenFlow is a protocol between SDN controllers and network devices, through which the controller instructs the OpenFlow switch to define the standard functional messages such as packet-received, send-packet-out, modify-forwarding table, and get-stats.

Besides ONF, many institutions have started the standardization work for SDN including international telecommunication union telecommunication standardization sector (ITU-T), European telecommunications standards institute (ETSI), China communications standards association (CCSA) and internet engineering task force (IETF). IETF issued an RFC concerning the requirements, application issues about SDN from operators' perspective [10], while ITU-T has no formal recommendation published since 2012 when the project was launched. Table 2.1 summarizes and lists all formally published SDN standards.

Currently, SDN has found its best practice in campus networks and data centers and is redefining the network architecture to support the new requirements of a new eco-system in the future. For example, Fig. 2.1 illustrates a generalized architecture of SDN controlled hybrid cellular networks. SDN techniques have been seen as promising enablers for future 5G HetNets, though necessary extensions are required.

Table 2.1 Current main SDN standards

Standardization organization	Main related standards and activities	Functionality
ONF	Software-defined networking: the new norm for networks (white paper) interoperability event technical paper v0.4/v1.0	Definition and interoperability
ITU	ITU-T resolution 77	Standardization for SDN
IETF	IETF RFC 7149	Perspective
ETSI	NFV ISG	Use cases and framework and requirements
CCSA	TC6 WG1	Application scenarios and framework protocol

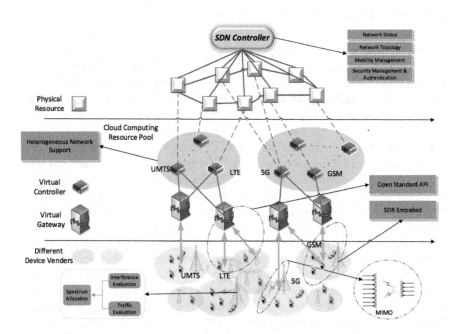

Fig. 2.1 A generalized architecture of SDN controlled hybrid cellular network

OpenFlow enables network controllers to determine the path of network packets across a network of switches. And it takes several years for the development of OpenFlow specification to become an academic initiative [11, 12]. In early 2011, ONF formalized the OpenFlow specification as the first standard communications interface between the control and forwarding layers of SDN architecture. ONF is funded by a lot of prominent companies such as Microsoft, Google, Yahoo, Facebook, Verizon, Deutsche Telekom and NTT. It is worth noting that these companies are all end users rather than network equipment manufacturers. This phenomenon indicates a strong market interest in OpenFlow technology. Furthermore, over 70 equipment manufacturers have joined the ONF, for example, Cisco, IBM, Brocade, and Juniper Networks. In June 2012, ONF released the latest specification version, i.e., OpenFlow 1.3, which is very important in datacenter and campus networking. So we must pay close attention to implementing only the approved industry standard version of these protocols to maximize interoperability in a multivendor network and fully realize the benefits intended by the ONF, as SDN and OpenFlow developers didn't stop innovating.

The OpenFlow standard integrates the network control plane into software running on an attached server or network controller, which enables the network control to become directly programmable and the underlying infrastructure to be abstracted for applications and network services. SDN concludes both industry standard network virtualization overlays and the emerging OpenFlow industry standard protocol. Most modern Ethernet switches and routers contain flow tables to match incoming packets to a particular flow and specify the functions that are to be performed on the packets. The flow tables run at line rate and are used to implement functions such as QoS, security firewalls, and statistical analysis of data streams. OpenFlow takes advantage of that to cope with network traffic involving a variety of protocols and network services. OpenFlow standardizes a common set of functions that operate on these flows and will be extended in the future as the standard evolves.

An OpenFlow switch is a software program or hardware device that forwards packets in the environment. It consists of three parts: (1) a remote controller on a server, (2) flow tables in the switch, and (3) a secure communication channel between them. A flow is a sequence of packets that matches a specific entry in a flow table, where the OpenFlow protocol allows external servers to define entries. In general terms, a flow could be all the packets from a particular MAC or IP address, a TCP connection, or all packets with the same virtual local area network (VLAN) identifier. Each flow table entry consists of a set of instructions that are executed if the packet matches the entry, for example, encapsulating and forwarding the flow to a controller for processing, forwarding the flow to a given switch port (at line rate), or dropping a flow packet.

As illustrated in Fig. 2.2, in the OpenFlow architecture the OpenFlow switch acts as a forwarding device containing one or more flow tables to manage the flows of packets through the switch. It communicates with SDN controller using the OpenFlow protocol running over the Secure Sockets Layer (SSL). Each flow table contains entries to determine how to process and forward packets belonging to a flow. Typically, flow entries consist of: (1) counters, updated for matching packets. The OpenFlow specification defines a variety of timers, such as number of received packets, number of bytes and duration of the flow; (2) a set of instructions, or actions, to be taken if a match occurs. They dictate how to handle matching packets; and (3) match fields, or matching rules, which may contain information found in the packet header, ingress port, and metadata, is used to select packets that match the values in the fields. Once a packet arrives at an OpenFlow switch, packet header

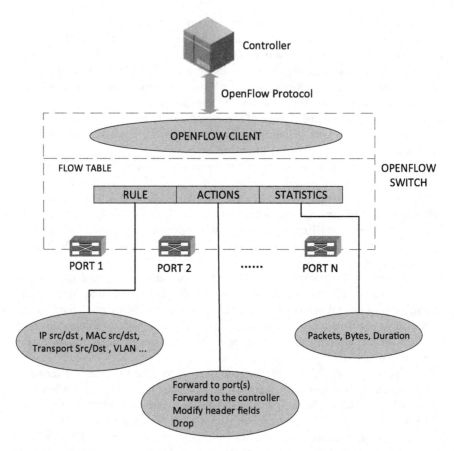

Fig. 2.2 Communication between the controller and the forwarding devices happens via OpenFlow protocol

fields are extracted and matched against the matching fields portion of the flow table entries. If there is a matching entry, the switch will execute appropriate set of instructions or actions associated with this entry. A flow table may also include a table-miss flow entry to instruct the action taken by the switch when there isn't a match during the flow table look-up procedure. In order to handle table misses, every flow table must contain a table-miss entry, which specifies a set of actions to be performed when there is no match for an incoming packet. On such condition, this entry may forward the packet to the controller over the OpenFlow channel to define a new flow or continue the matching process on the next flow table. If there is no match on any entry and there is no table-miss entry, then the packet is dropped.

It is important to note that OpenFlow supports multiple tables and pipeline processing from version 1.1. If there is more than one flow table in a switch, they are organized as a pipeline. Another possibility is to forward non-matching packets using regular IP forwarding schemes in the case of hybrid switches which have both OpenFlow and non-OpenFlow ports.

Overall, The OpenFlow protocol enables the controller to perform add, update, and delete actions to the flow entries in the flow tables reactively or proactively. OpenFlow protocol enables communication between controllers and switches by defining a set of messages which can be exchanged between these entries over a secure channel. OpenFlow improves scalability, enables multi-tenancy and resource pooling in cloud computing environments and will likely coexist with other layer 2/3 protocols and network overlays for some time.

As illustrated in Fig. 2.3, in SDN architecture, the data center network is virtualized with a software overlay and network controller which allows attached servers to control features such as packet flows, topology changes, and network management [13, 14].

OpenFlow architecture, as shown in Fig. 2.4, consists of three main components: an OpenFlow-compliant switch, a secure channel and a controller. Switches forward packets by means of flow tables, which tell the switch how to process each data flow by associating an action with each flow table entry. The flow table entries consist of counters, instructions and match fields. If the incoming packets match the

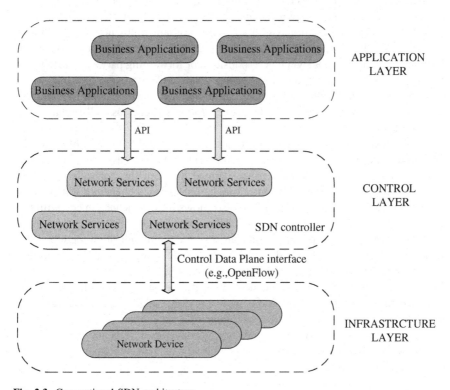

Fig. 2.3 Conventional SDN architecture

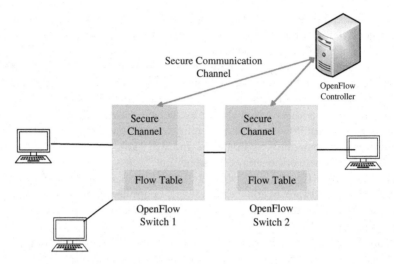

Fig. 2.4 OpenFlow architecture

match fields of an entry, the packets will be processed according to the action contained by that entry. When there is not a match, it is also possible to send the packet to the controller. Counters are responsible for keeping statistics about packets [15, 16]. The controller is a software program following the OpenFlow protocol, which is the responsible for managing the OpenFlow switches. The OpenFlow-compliant switch is responsible for forwarding packets according to the rules defined in the flow table. The secure channel Connects the controller to all the switches, so commands and packets can be sent between the controller and the switch, such as receiving and sending packets to the switches as well as managing these switches.

OpenFlow has been one of the most commonly deployed SDN technologies [17] for implementing SDN in networking equipment. ONF defines OpenFlow as the first standard communications interface defined between the controller and forwarding layers of an SDN architecture, which enables researchers to test new ideas in a production environment. OpenFlow also shows its significance in supporting a standard, secure protocol between the SDN controller and the network device. In SDN architecture, OpenFlow standardizes the communication between the switches and the software-based controller [18, 19]. It allows switches from different vendors to be managed remotely using a single, open protocol and doesn't require the vendors to expose the code of their devices when controlling a switch. The original intention of OpenFlow was to provide a platform that would allow researchers to run experiments to decide how data packets are forwarded through the production networks. However, industry has developed SDN and OpenFlow as a strategy to reduce costs and hardware complexity, as well as increasing the value of data center services.

OpenFlow standardizes a common set of functions in OpenFlow networks. For example, it is possible to control multiple switches from a single controller and to analyze traffic statistics in OpenFlow networks using software. It is also feasible to update forwarding information dynamically and exchange *streams of packets* between server and client which can be abstractly as flows. Research communities have exploited these capabilities to experiment with innovative ideas and propose new applications. In the study of OpenFlow, most research focus on security, availability, network management, network and data center virtualization, wireless applications and ease of configuration, which have been implemented in different environments including both real hardware networks and virtual simulations [20, 21].

Many specific capabilities of OpenFlow-based architectures such as centralized control, flow abstraction, dynamic updating of forwarding rules and software-based traffic analysis are worthy of exploitation to experiment with new ideas and test novel applications. Implementing OpenFlow can provide integrated network management and ease the configuration of a network, as well as adding security features. Openflow also assists in virtualizing networks and data centers and deploying mobile systems. These applications run on top of networking operating systems such as Node.Flow, Floodlight, Beacon, Maestro, Trema and Nox. Many investigations choose to measure the performance of OpenFlow networks through experimentation and modeling. The research communities can run experiments and test their applications in a more realistic scenario by deploying larger scale OpenFlow infrastructures.

In the following, we elaborate the details of dynamic QoS support as an example of the advantage of SDN paradigm for the 5G core network.

OpenFlow is a new method of control for flows in the network, which offers a new paradigm to make routing more flexible by allowing different routing rules associated with data flows. It defines a standard for sending flow rules to network devices so that the modification of partitions of the networks layout and traffic flows can be as instant as in an SDN. The controller component of OpenFlow is where routing changes are determined. It acts as the brain of the network and handles the incoming data packets. Thus, different data flows associated with different algorithms in the controller may yield different routing choices. By adding the flow rules to the existing forwarding table in the network device, the controller provides access to flow tables as well as the rules which instruct the network forwarders how to direct traffic flows. As illustrated in Fig. 2.5, the QoS supporting controller offers various functions and interfaces, and some of them have been part of a router in the classical Internet model.

Controller-Forwarder interface and Controller-Application interface are two main interfaces of a controller. At Controller-Forwarder interface, the controller attaches to the forwarders through a secure OpenFlow protocol. This interface is responsible for discovering network topology by requesting network state information from forwarders, sending flow tables associated with data flows, and receiving traffic status information and notifications. At Controller-Application interface, the controller provides a secure and open (non-real time) interface for application service providers to make reservations of new data partitions (individual or groups of data flows).

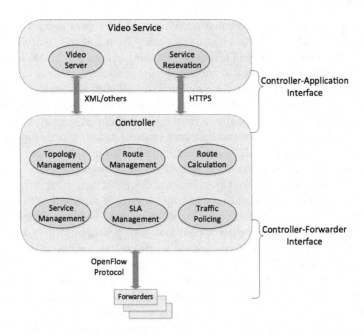

Fig. 2.5 The OpenFlow controller and interfaces

When necessary, it can also define new routing rules associated with partitions. There are several key functions under the management of the controller.

Route management: Route management is responsible for analyzing availability and packet performance of routes to aid the route calculation. It requires collecting performance data from the forwarders on a periodic or on a notification basis.

Route calculation: This function is used for finding and calculating routes for different data flows. Several routing algorithms can run in parallel to meet the performance requirements and objectives of different flows. This function contains service reservations along with network topology and route management information.

Topology management: This function is responsible for determining and maintaining network connectivity topology by analyzing data collected or received from forwarders.

Service level agreement (SLA) management: SLA provides users SLA templates for different types of allowable data flow patterns.

Traffic policing: this function determines whether data flows agree with their SLAs, and apply policy rules when they do not.

Service management: This function is responsible for managing and storing different data flow request.

The process of the service management function of controller is as follows:

(1) The controller receives a service reservation module (e.g. video service) requiring QoS option from the network.

(2) The controller decides if other reservations being made can deliver the requested service SLA.

(3) Once accepting the reservation, the controller manages the network resources for the service starting at the begin time of the data flow.

(4) The route calculation function computes the exact route and uploads the flow table to appropriate forwarders and the video server signals the controller and data flow may start.

(5) The route management function reactivates the route calculation function to determine a new route for the data flow if there is congestion in the network.

It is worth noting that, in order to make sure the end points conform to their SLAs stated in their QoS contract, the traffic policing function must also be implemented.

2.3 Key Technologies in 5G HetNets

In the era of 5G, a major challenge is to efficiently support the increasing demand of network capacity while the spectrum resource remains scarce [22]. For example, the future networks is expected to be able to handle the sophisticated content operations of a tenfold increase in traffic with guaranteed quality of service (QoS) [23, 24]. Additionally, the future networks also requires high energy efficiency, since service providers are more and more concerned about the OPEX, as well as the impact to global climate change and air pollution [25, 26].

To address these challenges, operators have three plans to cope with the current network:

(1) Deploying denser cellular infrastructures.
(2) Increasing the available bandwidth.
(3) Exploring massive MIMO solutions [27, 28].

In particular, one of the most promising technologies is the HetNet architecture, where a cellular system consists of a large number of densified low power nodes [29, 30]. LPNs can improve the system capacity with frequency reuse and provide high data rate to nearby mobile stations In HetNets. Additionally, LPNs can transmit signals with lower power to achieve a significant reduction in energy consumption [31, 32].

Nevertheless, the total energy consumption may still likely exceed the acceptable level despite the promising features of massive MIMO based HetNets and the drastic increase of ultra-dense deployment of small cells in 5G networks [33]. Moreover, the traffic fluctuation contributes considerable increase to energy consumption and thus causes extra OPEX to service providers [34]. Thus, wireless network operators should introduce intelligence to deal with the rapid growth of users and traffics [35, 36].

Due to the ever increasing network complexity, the operators are driven toward a virtualization of network functionality that calls for a paradigm shift from a hardware-based approach to a software-based approach [37, 38]. We will correspondingly develop an intelligent management framework based on the concept of SDN, which is featured by the decoupling of control plane from data plane [39, 40]. The intelligent SDN framework aims to provide a viable way to solve the existing challenges in a unified manner [41, 42].

Essentially, SDN recognizes the network as an operating system and abstracts the applications from the hardware [43]. It enables the management-related functions to be implemented in a centralized manner [44]. In this way, network intelligence can be realized logically in a centralized SDN controller that manages the entire network globally. Concretely, an SDN controller is a programmable device which can learn the physical topology of the network and status of each individual network element through certain discovery mechanisms or appropriate databases. Thus it can orchestrate the whole network to function in a cost-efficient and energy-efficient manner [45].

2.3.1   Heterogeneous Networks

Heterogeneous networks are now an established concept within LTE networks. HetNets have been considered as the most viable solution to the impending mobile data traffic crunch in the context of LTE-A. It is comprised of a combination of different access technologies and different cell layers, such as macro, micro, pico and femto. HetNets can significantly increase the spectral efficiency by manipulating these layers in an appropriate manner [46, 47]. In HetNets, Macro cells are used to provide coverage. Pico cells and micro cells are used to enhance capacity in busy areas, such as train stations, shopping malls and city centers. HetNets can also enhance the key requirements for disaster rescue scenarios of the network, such as availability, reliability and survivability. HetNets take full advantage of the complementary characteristics of different network tiers, and thus become an inevitable trend for future development of information networks [48].

Though there are many benefits in deploying HetNets, it causes a new problem we have to handle, i.e., spatial domain management, which is due to various service requirements and different tiers of networking access technologies in HetNets [49, 50]. Fortunately, in 5G HetNets, we can rely on SDN controller to collect and manage spatial information from a variety of high and low power nodes.

Another technology, small cell network, which has a deployment of low power nodes, has been considered as an important technology of next generation cellular networks as shown in Fig. 2.6. It can improve spectrum efficiency, power efficiency, and coverage effectively, as well as reduce CAPEX and OPEX [51, 52].

To enable the increasing quality experience for demanding smartphone users, mobile networks have to continuously evolve to meet capacity and coverage demands with the latest technologies. Meanwhile, although the recent mobile data

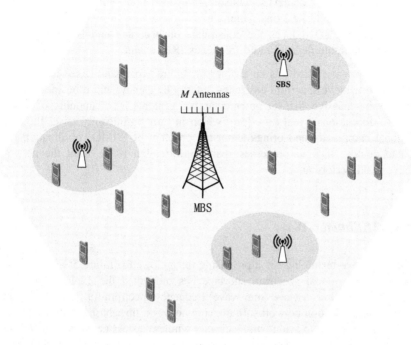

Fig. 2.6 A small cell network

usage has an exponential growth, there is an apparent trend of declining profitability of mobile data. Moreover, dense small cell networks have attracted great attentions, which deploy a mass of small cell base stations (SBSs) to improve QoS further [53].

2.3.2 Massive MIMO

Massive MIMO (also known as Large-Scale Antenna Systems and Very Large MIMO) is becoming mature for wireless communications. It employs the antenna arrays of a few hundred elements to simultaneously serve many tens of mobile users with the same time-frequency resource [54, 55]. Basically, the more antennas the transmitter/receiver is equipped with, the more the possible signal paths and the better the performance in terms of data rate and link reliability. Massive MIMO scales up the conventional MIMO to reap all the benefits at a higher level:

(1) Massive MIMO increases the capacity times or more and simultaneously improving the radiated energy efficiency in the order of 100 times;
(2) Massive MIMO applies inexpensive, low-power components;
(3) Massive MIMO gains a significant reduction of latency on the air interface;

(4) Massive MIMO achieves robustness against both unintended man-made interference and intentional jamming;

(5) Massive MIMO reduces the constraints on accuracy and linearity of each individual amplifier and radio frequency (RF) chain.

Overall, massive MIMO is an enabler for future broadband (fixed and mobile) networks to achieve spectrum efficiency, energy efficiency, and robustness.

However, massive MIMO antenna arrays generate vast amounts of spatial domain information in real time [56], which in turn significantly raises the computational complexity and brings about the problem of MIMO coordination. We need to deploy SDN to processes spatial information if we want the network achieve good performance.

2.3.3 Millimeter-Wave

Millimeter-wave technology is a promising technology for future 5G cellular systems [57]. The global spectrum shortage has motivated the exploration of the underutilized millimeter wave (mm-wave) frequency spectrum for future broadband cellular communication networks. In the core network, fiber based millimeter-wave systems are expected to bridge high capacity wireless access networks in the future [58]. Particularly, millimeter-wave signals on optical carriers are up and down linked between base antenna stations and central stations. Future 5G cellular systems can achieve super wide bandwidth from millimeter-wave, as its frequency ranges from 26.5 to 300 GHz. In radio access network, millimeter-wave has an operating frequency between microwave and light and owns the advantages of both. Furthermore, millimeter-wave has a much narrower antenna beam size compared to the microwave, so that it can more precisely aim the target [59].

In 5G network, millimeter-wave has brought the challenge of acquiring transmission characteristic in the air and the requirement of high device precision. Spatial domain management makes the control of millimeter-wave devices more efficient without the need to install hardware for every new service, and thus offers a chance of low equipment cost and operational cost [60, 61].

2.4 Intelligent SDN Architecture for 5G HetNets

Future 5G system will employ more effective technologies to support the interconnection of more diversified user equipment (UE) and devices, including traffic control, resource allocation, density management, security, etc. The existing SDN standards mainly address the management of wired network such as 5G core network. However, there still exist a large number of wireless technologies as another essential part in radio access network (RAN) uncovered. In the following, we will extend the function of SDN to involve the 5G wireless technologies, such as HetNets, massive MIMO, etc. We will emphasize the function of SDN with respect to radio resource management in massive MIMO HetNets.

We have to exploit SDN technology to empower the network with intellectualization, so as to lower the cost on network deployment/maintenance, reduce the complexity of 5G networks, and facilitate the future network evolution. However, the current standardizations have not taken the 5G RAN into consideration. Next, we will propose the necessary extensions to current standards and our novel SDN architecture. Particularly, we will highlight the problem of the huge spatial domain information management and complicated MIMO coordination.

Figure 2.7 demonstrates our proposed architecture, where the 5G networks employ an SDN controller to conduct massive MIMO coordination and manage spatial information. In this architecture, the network consists of BS as well as densely deployed small cells supporting spectral efficiency and coverage. Additionally, Both BS and LPNs can be extended to have a huge number of antennas, i.e. massive MIMO. Here SDN controller overcomes the massive data processing constraints by taking responsibility of radio resource management (RRM) of the whole HetNet. In the RRM procedure, BS collects the user information and report it to SDN. Then SDN controller will perform the information processing including CSI analysis and null-space calculation. After that, SDN controller sends the processed results and instructions back to BS and LPNs, and thus they can achieve better coordination regarding the global knowledge of

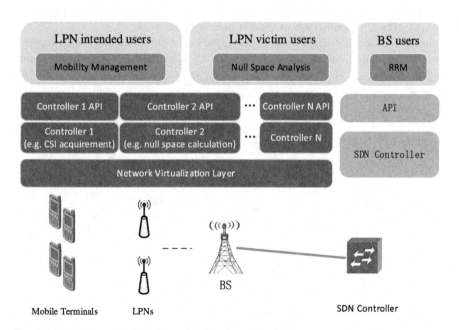

Fig. 2.7 5G SDN architecture for spatial domain management

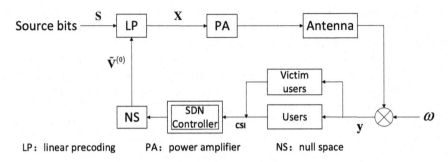

Fig. 2.8 Block diagram of our proposed scheme

network spatial information. In our proposed architecture, SDN takes over the heavy operation from BS and LPNs to achieve a better efficiency for 5G networks.

Next, we will present a novel SDN architecture for 5G networks and our proposed scheme to implement massive MIMO coordination in 5G SDN. Figure 2.8 illustrates the proposed processing diagram, where we highlights two critical steps, the acquisition of CSI for LPNs and the generation of precoding matrix based on the null-space of the victim users. The details will be elaborated in the following parts.

In conventional architectures, LPNs can only collect the local CSI of their own users though the backward channel and cannot access to the CSI of the external victim users. Fortunately, the CSI of all MIMO channels can be collected and disseminated through the backhaul link owing to the hub spoke structure of the network. The CSI acquisition method is demonstrated in Fig. 2.9.

In the CSI acquisition procedure, BS collects user information and then reports the channel matrix on victim users of each LPN to the SDN controller. More specifically, BS sends SDN every $[\mathbf{H}^H_{j,Kj+1}, \ldots, \mathbf{H}^H_{j,Kj+Lj}]^H$ corresponding to the jth LPN, which contains $L_j \times M$ elements. Then SDN controller will perform the block-diagonal (BD) algorithm and derive null-space vector $\tilde{\mathbf{V}}^{(0)}_{j,i}$ for the corresponding LPN j. Next, SDN controller computes on the null-space vector $\tilde{\mathbf{V}}^{(0)}_{j,i}$ to

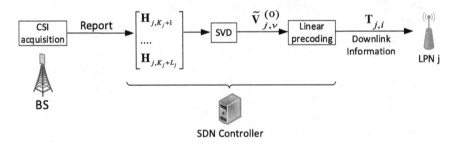

Fig. 2.9 Proposed CSI acquirement methods

obtain precoding matrix $\mathbf{T}_{j,i}$ ($i = 1, 2\ldots K_j$) by the precoding algorithm as proposed later. Finally, it sends $\mathbf{T}_{j,i}$ to each LPN through downlink transmission. In this way, the SDN controller takes over all the matrix decomposition computation and precoding methods and it is suitable for the situation that LPNs are simple and devices are in low performance.

2.5   NFV Technology Overview

Due to the explosion of mixed utilization of highly diversified access technologies, the emerging 4G/5G wireless networks have been characterized by heterogeneity. There are several challenges for 5G designers such as proposing cost-efficient and energy-saving solutions. In the following, we will introduce a promising technology for 5G wireless communication systems, namely network function virtualization (NFV).

In computer science, network function virtualization (NFV) is a network architecture concept that uses the technologies of IT virtualization to virtualize entire classes of network node functions into building blocks that may connect, or chain together, to create communication services. In this way, the operators can architect their networks towards deploying network services onto these standard devices [62]. NFV decouples the network functions, such as network address translation (NAT), firewalling, intrusion detection, domain name service (DNS), and caching, to name a few, from proprietary hardware appliances so they can run in software, and thus offers a chance of less investment in network equipment and expenditure on network management and operation. It's designed to consolidate and deliver the networking components needed to support a fully virtualized infrastructure, which exactly meets the mandate of cloud computing environment.

NFV is closely related to other two technologies, i.e., software defined radio (SDR) and SDN when taking part in 5G. Figure 2.10 shows the evolution of the broadband radio access technologies (RAT) along with the key characteristics, such as cell sizes, service bearer requirement and frequency resources. From 2G to 5G, not only data throughput and mobility increases but also network structure changes. Apparently, the cell sizes shrink continuously to achieve better coverage and network capacity. Moreover, the pivotal service has shifted from circuit switched voice to packet switched broadband data access and multimedia telephony. This trend requires the augmentation of network intelligence with more swift and cost efficient resource provisioning for specific applications guaranteeing QoS. In the era of 5G, the network must be capable of meeting a huge amount of user diversified service demands at different data rates. It, therefore, should be intelligent enough to orchestrate all the available frequency resources, infrastructures and hardware devices. The industry proposed the concept of HetNet as a hierarchical network resource utilization scheme exploiting both cellular and WLAN technologies. NFV, SDR and SDN can support HetNet at different levels. For the service carrier, the programmable NFV devices can instantiate and schedule different service systems

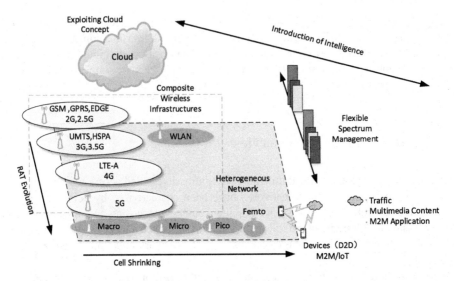

Fig. 2.10 Evolution of mobile network and key technologies

to meet various requirements and setting up new networks; for the interfaces of heterogeneous nodes and devices, SDR extracts data packets for upper layers and provides the physical layer signal processing; for end-to-end packet transportation, SDN can flexibly construct the end-to-end congestion resilient transmission channels by exploiting various the network protocols.

Figure 2.11 illustrates a hybrid architecture of NFV, SDR and SDN. NFV technology is a new method to build an end-to-end network infrastructure with evolving

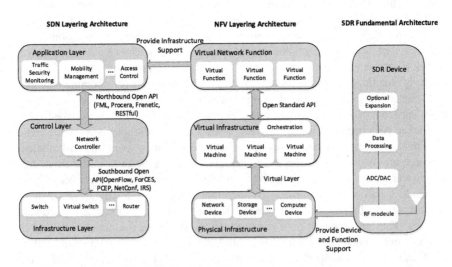

Fig. 2.11 Architecture of NFV, SDR and SDN

standard IT virtualization technology that runs on high-volume service, switch and storage hardware to virtualize network functions. This can be achieved by means of implementing network functions via software. As a result, it can avoid specific hardware-based devices and increase the return on investment. It also reduces the time of deploying new networking services to seize new market opportunities and support changing business requirements. SDR replaces the hardware implementation with the software on a personal computer or embedded system. The receiver also can use a computer to recover the signal with intelligence. SDR can be flexible enough to avoid the "limited spectrum" assumptions of designers of previous kinds of radios. It overcomes the disadvantages of radio hardware. SDN can help to split the control functionality and the service functionality of the network, which is considered to be beneficial to the development of the future network [63].

The goal of NFV is to decouple network functions from dedicated hardware devices and allow network services that are now being carried out by routers, firewalls, load balancers and other dedicated hardware devices to be hosted on virtual machines (VMs). We believe that NFV is applicable to any data plane packet processing and control plane function in fixed and mobile network infrastructures. In NFV technology, the network function of a network device is implemented in a software package running in virtual machines. NFV enables the establishment of new mechanisms to deploy and operate network and infrastructure services, Also lowers the risks associated with rolling out new services, allowing providers to easily trial and evolve services to determine what best meets the needs of customers. It can be rapidly provisioned and released with minimal management effort or service provider interaction [64, 65]. Figure 2.11 shows the layering architecture of NFV, which focuses on the functionalities necessary for the virtualization and the consequent operation of an operator's network. The blocks here including virtual network function, virtual infrastructure and physical infrastructure, are the potential targets for standardization. The open standard API is available in present deployments but might need extension for handing network function virtualization.

SDR attempts to place much or most of the complex signal handling involved in communications receivers and transmitters into the digital (DSP) style. The ideal SDR has significant utility for the military and cell phone services, because it has multiple bands and multiple modes with open architecture. Its wireless functions are achieved by loading the software offering a variety of radio communication services. A basic SDR system may consist of antenna, multi-band radio frequency (RF) module, broadband A/D (D/A) converter, DSP processors and other expansions if possible. SDR moves A/D and D/A closer to the RF, from the base to the intermediate frequency or even to RF, and replaces dedicated digital circuit by programmable DSP or FPGA devices [66]. By separating the system hardware architecture and function this way, SDR can achieve the flexibility and low cost, as well as other wide-reaching benefits realized by service providers and product developers through to end users. It can also program the operating frequency, system bandwidth, modulation and source encoding [67]. Moreover, the system update can be easily done by changing the process modules according to the requirements.

SDN is a new network architecture, which separates the data and control functions of networking devices, such as routers, packet switches, and LAN switches, with a well-defined Application Programming Interface (API) between the two. This migration of control enables the underlying infrastructure to be abstracted for applications and network services, which can treat the network as a logical or virtual entity [68]. It allows networks to react dynamically to changes in usage patterns and availability of network resources. SDN architecture is shown in the left part of Fig. 2.11. Network intelligence is (logically) centralized in software-based SDN controllers, making it possible for networks to interact with applications and efficiently reconfigure themselves at need, allowing them to implement multiple logical network topologies on a single common network fabric. With SDN, enterprises and carriers can achieve vendor-independent control over the entire network from a single logical point, which greatly simplifies the network design and operation. Also, owning to SDN property, the network devices no longer need to understand and process thousands of protocol standards but merely accept instructions from the SDN controllers. Thus the network can be simplified. With Southbound Open APIs (OpenFlow, ForCES, PCEP, NetConf, IRS) between the SDN control and infrastructure layers, it allows direct manipulation of the forwarding plane of network devices, such as switches and routers, both physically and virtually (hypervisor-based). Additionally, business applications, such as mobility management, traffic security monitoring and access control, can operate on an abstraction of the network with Northbound Open APIs (FML, Procera, Frenetic, RESTful) between the SDN control and application layers, leveraging network services and capabilities without being tied to the details of their implementation [69]. SDN makes the network not so much application-aware as application-customized and applications not so much network-aware as network-capability-aware. As a result, it optimized the computing, storage, and resources in the networks.

In Fig. 2.11, NFV can increase the flexibility of network service deployment in an operator's network. SDN can assist NFV to implement these functions in terms of enhancing performance, simplifying compatibility with existing deployments and facilitating operation and maintenance procedures. Practically NFV aligns closely to the SDN objectives by using commodity servers and switches, and thus administrators can easily collect the gathered results and ultimately move forward to a self-adaptive environment.

2.6   Necessary Standard Extensions for Enabling 5G

As aforementioned, in order to support the interconnection of more diversified user equipment and devices, the future 5G wireless networks are characterized by the heterogeneity and will employ more sophisticated technologies for spectrum utilization, multi-access, signal processing, security etc. The advancement of NFV, SDR and SDN is the key to evolving the network to keep pace with the innovations of all the people and devices its connecting. However, the current standardization work bodies have not taken the 5G network into consideration. In the following, we will present the necessary extensions to current standards.

From the network operation perspective, these three technologies concern different layers of a network. Specifically, SDR takes charge of physical layer issues such as modulation, signal process, channel coding and data encapsulation; SDN focuses on switching and routing of data packets with guaranteed QoS, as well as scheduling network elements to ensure the end to end transportation; NFV orchestrates and combines the functional entities of the network to ensure a quickly deployed network.

The application of MIMO technologies and the dynamic spectrum utilization enables the high data throughput and mobility in 5G network. To achieve these enhancements, the technologies need to be embedded in SDR. The emerging technologies multiuser MIMO and network MIMO can improve network access efficiency by shaping and exploiting interference, and are considered as candidate technologies for 5G. However, the high performance of the new paradigm is obtained at the expense of significantly increased computation complexity. Moreover, As for 5G, the heterogeneous network structure shall ensure that the devices must be programmed to adapted to the underlining network schemes, such as WiFi or cellular, making full use of spectrum resources either licensed or not. To achieve the functions above, more effective architecture and efficient algorithms need to be defined and introduced to the SDR framework, such as parallel data processing and software enabled power saving modes.

SDN has practiced well in campus networks and datacenters. As to 5G networks, the challenging applications mainly lie in mobile management, inter-ISP handoff and the control plane security. Since SDN centralizes control of the network by separating the control logic to off-device computer resources, the switching of multi-homing user equipment between heterogeneous networks, such as Wi-Fi and cellular, can be easily managed purely at higher level without sudden loss of connectivity or service interruption. The newer standards such as IEEE 802.11f, 802.11k and 802.11r has make amendments to tackle this kind of homogeneous handover situations and the extension to SDN can be straightforward. To achieve seamless handoff between the networks of different service providers, IETF proposed Mobile IP (MIP) to allow roaming devices to move from one network to another with a permanent IP address attached. And it can be supported by SDN in a more efficient way. Practically, the OpenFlow controller of different ISP can communicate with one another to maintain and update a roaming IP-address table globally. In 5G networks, this kind of efficiency improvement exploited through cross-layer optimization can find even more application scenarios. Moreover, future SDN standards should emphasize on security. More specifically, when the session extends from private networks to public ones of ISP, we should pay attention to the protection of subscriber privacy, robust mechanisms to cyber and malware attacks, etc.

As for NFV, the standard enrichment should be based on network operator's perspective:

(1) The evolution of NFV must conform to the trend of wireless network development;
(2) The issues of migration and co-existence of present legacy networks and NFV enabled ones need to be considered thoroughly;
(3) NFV should be a healthy ecosystem to support the sustainable development.

The mixture of infrastructure-based and infrastructure-less networks will be the trend for 5G. Moreover, in 5G network, the interconnection will not be limited to people only but also among things and machines. Accordingly, NFV standards should extend to support a hybrid network with extended abilities, such as self-organizing. At the same time, it needs to support different kinds of devices to enable a variety of new applications such as vehicular communication, health care delivery and environmental monitoring. The coexistence issue concerns about the interoperability of the legacy and new virtualized networks and compatibility of business and operation supporting systems (BOSS). The NFV architecture needs to enable a well functioned hybrid network supporting both classical physical network appliances and virtual network appliances. It should also provide a smooth path of migration from the classical network to an open standard based virtual one.

Furthermore, NFV should enable overlapped functions with traditional BOSS of classical networks. There is a large possibility that the BOSS functions are also virtualized and merged into NFV orchestration, which requires the cooperation with BSS and OSS standardization bodies such as TMForum. The interoperability and portability of virtual appliances for network equipment will contribute to a healthy ecosystem of NFV. Portability shall bring the freedom to optimize service and network deployment and the interoperability decouples the virtual appliances from

Table 2.2 Necessary standard extensions for enabling 5G

Extension for 5G	Challenges
NFV	
Open standard API	Define a unified interface, which clearly decouples the software instances from the underlying hardware, as represented by virtual machines and their hypervisors
Embedded SDR	Define the standard to bridge SDR device and unified underlying NFV hardware
SDR	
Multi-input multi-output (MIMO)	Attains higher spectrum efficiency at the cost of significantly increased computation complexity
Radio resource management	Increase the system complexity with the still unknown power management and device mode configuration in SDR
SDN	
Mobility management	Seamless handoff between service providers
Heterogeneous network support	Provide indiscriminated service regardless of location or type of network access
Security	Overhead of SDN control and privacy issues

the physical equipment provided by different vendors. These functions need NFV ISG to define unified interfaces and protocol to decouple the abstracted function from the underlying hardware, which is similar to OpenFlow. Table 2.2 summarizes and lists the suggested extensions to the standards of three technologies.

References

1. C.-X. Wang et al., Cellular architecture and key technologies for 5G wireless communication networks. IEEE Commun. Mag. **52**(2), 122–130 (2014)
2. V. Chandrasekhar, J.G. Andrews, A. Gatherer, Femtocell networks: a survey. IEEE Commun. Mag. **46**(9), 59–67 (2008)
3. F. Rusek et al., Scaling up MIMO: opportunities and challenges with very large arrays. IEEE Sign. Process. Mag. **30**(1), 40–60 (2013)
4. A. Bleicher, Millimeter waves may be the future of 5G phones. Samsung's millimeter-wave transceiver technology could enable ultrafast mobile broadband by 2020, June 2013
5. H. Haas, Wireless data from every light bulb. (Aug 2011), http://bit.ly/tedvlc
6. X. Hong, C.-X. Wang, H.-H. Chen, Y. Zhang, Secondary spectrum access networks. IEEE Veh. Technol. Mag. **4**(2), 36–43 (2009)
7. F. Haider et al., in *Spectral Efficiency Analysis of Mobile Femtocell Based Cellular Systems.* Proceedings of IEEE ICCT, Jinan, China, Sept 2011, pp. 347–351
8. P. Agyapong, M. Iwamura, D. Staehle, W. Kiess, A. Benjebbour, Design considerations for a 5G network architecture. IEEE Commun. Mag. **52**(11), 65–75 (2014)
9. C. Rotsos, N. Sarrar, S. Uhlig, R. Sherwood, A.W. Moore, in *OFLOPS: An Open Framework for OpenFlow Switch Evaluation.* Passive and Active Measurement, pp. 85–95, Jan 2012
10. J.H. Jafarian, E. Al-Shaer, Q. Duan, S. Murakami, in *Openflow Random Host Mutation: Transparent Moving Target Defense Using Software Defined Networking.* Proceedings of the First Workshop on HOT Topics in Software Defined Networks, pp. 127–132, Aug 2012
11. X. Xu, H. Zhang, X. Dai, Y. Hou, X. Tao, P. Zhang, SDN based next generation mobile network with service slicing and trials. Commun. China **11**(2), 65–77 (2014)
12. J. Sanchez, I.G. Ben Yahia, N. Crespi, T. Rasheed, D. Siracusa, in *Softwarized 5G Networks Resiliency with Self-Healing.* 5G for Ubiquitous Connectivity (5GU), pp. 229–233, Nov 2014
13. X. Li, H. Zhang, in *Creating Logical Zones FOR Hierarchical Traffic Engineering Optimization in SDN-Empowered 5G.* Computing, Networking and Communications (ICNC), pp. 1071–1075, Feb 2015
14. X. Duan, X. Wang, Authentication handover and privacy protection in 5G hetnets using software-defined networking. Commun. Mag. IEEE **53**(4), 28–35 (2015)
15. S. Kuklinski, Y. Li, K.T. Dinh, in *Handover Management in SDN-Based Mobile Networks.* Globecom Workshops (GC Wkshps), pp. 194–200, Dec 2014
16. V. Jungnickel, K. Habel, M. Parker, S. Walker, C. Bock, J. Ferrer Riera, V. Marques, D. Levi, in *Software-Defined Open Architecture for Front and Backhaul in 5G Mobile Networks.* Transparent Optical Networks (ICTON), pp. 1–4, July 2014
17. L. Liu, R. Muoz, R. Casellas, T. Tsuritani, R. Marthez, I. Morita, OpenSlice: an openflow-based control plane for spectrum sliced elastic optical path networks. Opt. Express **21**(4), 4194–4204 (2013)
18. A. Lara, A. Kolasani, B. Ramamurthy, Network innovation using openflow: a survey. Opt. Express Commun. Surv. Tutorials **16**(1), 493–512 (2014)

19. L. Liu, D. Zhang, T. Tsuritani, R. Vilalta, R. Casellas, L. Hong, in *First Field Trial of an OpenFlow-Based Unified Control Plane For Multilayer Multi-Granularity Optical Networks*. National Fiber Optic Engineers Conference, pp. PDP5D-2, Mar 2012

20. T. Luo, H.P. Tan, T.Q. Quek, Sensor OpenFlow: enabling software defined wireless sensor networks. Commun. Lett. IEEE **16**(11), 1896–1899 (2012)

21. M. Channegowda, R. Nejabati, M. Rashidi Fard, S. Peng, N. Amaya, G. Zervas, Experimental demonstration of an OpenFlow based software defined optical network employing packet, fixed and flexible DWDM grid technologies on an international multi-domain testbed. Opt. Express **21**(5), 5487–5498 (2013)

22. R.Q. Hu, Y. Qian, An energy efficient and spectrum efficient wireless heterogeneous network framework for 5G systems. IEEE Commun. Mag. **52**(5), 94–101, May 2014

23. M. Palkovic, P. Raghavan, M. Li, A. Dejonghe, L. Van der Perre, F. Catthoor, Future software-defined radio platforms and mapping flows. IEEE Sig. Process Mag. **23**(4), 22–33 (2010)

24. Y. Xu, R.Q. Hu, L. Wei, G. Wu, in *QoE-Aware Mobile Association and Resource Allocation Over Wireless Heterogeneous Networks*. Global Communications Conference (GLOBECOM), pp. 4695–4701, Dec 2014

25. I. Chih-Lin, C. Rowell, S. Han, Z. Xu, G. Li, Z. Pan, Toward green and soft: a 5G perspective. IEEE Commun. Mag. **52**(2), 66–73 (2014)

26. R.Q. Hu, Y. Qian, S. Kota, G. Giambene, HetNets—a new paradigm for increasing cellular capacity and coverage. IEEE Trans. Wireless Commun. **18**(3), 8–9 (2011)

27. R.Q. Hu, Y. Qian, *Heterogeneous Cellular Networks* (Wiley, London, 2013)

28. R.L.G. Cavalcante, S. Stanczak, M. Schubert, A. Eisenblaetter, U. Tuerke, Toward energy-efficient 5G wireless communications technologies: tools for decoupling the scaling of networks from the growth of operating power. IEEE Sig. Process. Mag. **31**(6), 24–34 (2014)

29. H. Masutani, Y. Nakajima, T. Kinoshita, T. Hibi, H. Takahashi, K. Obana, K. Shimano, M. Fukui, in *Requirements and Design of Flexible NFV Network Infrastructure Node Leveraging SDN/OpenFlow*. IEEE Optical Network Design and Modeling, pp. 258–263, May 2014

30. B.A.A. Nunes, M. Mendonca, X.-N. Nguyen, K. Obraczka, T. Turletti, A survey of software-defined networking: past, present, and future of programmable networks. IEEE Commun. Sur. Tutorials **16**(3), 1617–1634 (2014)

31. X. Duan, X. Wang, Authentication handover and privacy protection in 5G hetnets using software-defined networking. IEEE Commun. Mag. **53**(4), 28–35 (2015)

32. S. Sun, M. Kadoch, L. Gong, B. Rong, Integrating network function virtualization with SDR and SDN for 4G/5G networks. Netw. IEEE **29**(3), 54–59 (2015)

33. S. Sun, B. Rong, Y. Qian, Artificial frequency selective channel for covert CDD-OFDM transmission. J. Secur. Commun. Netw. (2014)

34. Q. Li, R.Q. Hu, Y. Qian, G. Wu, Cooperative communications for wireless networks: Techniques and applications in LTE-advanced systems. IEEE Trans. Wireless Commun. **19** (2), 22–29 (2012)

35. D. Choudhury, in *5G Wireless and Millimeter Wave Technology Evolution: An Overview*. IEEE MTT-S International Microwave Symposium (IMS), pp. 1–4, May 2015

36. S. Han, I. Chih-Lin, Z. Xu, C. Rowell, Large-scale antenna systems with hybrid analog and digital beamforming for millimeter wave 5G. IEEE Commun. Mag. **53**(1), 186–194 (2015)

37. V. Jungnickel, K. Manolakis, W. Zirwas, B. Panzner, V. Braun, M Ossow, M. Sternad, R. Apelfrojd, T. Svensson, The role of small cells, coordinated multipoint, and massive MIMO in 5G. IEEE Commun. Mag. **52**(5), 44–51 (2014)

38. B. Panzner, W. Zirwas, S. Dierks, M. Lauridsen, P. Mogensen, K. Pajukoski, D. Miao, in *Deployment and Implementation Strategies for Massive MIMO in 5G*. Globecom Workshops (GC Wkshps), 2014, pp. 346–351, Dec 2014

39. T.R. Omar, A.E. Kamal, J.M. Chang, in *Downlink Spectrum Allocation in 5G HetNets*. Wireless Communications and Mobile Computing Conference (IWCMC), pp. 12–17, Aug 2014
40. D. Liu, L. Wang, Y. Chen, T. Zhang, K. Chai, M. Elkashlan, Distributed energy efficient fair user association in Massive MIMO enabled HetNets. IEEE Commun. Lett. **99**, 1–1 (2015)
41. C.-F. Lai, R.-H. Hwang, H.-C. Chao, M. Hassan, A. Alamri, A buffer-aware HTTP live streaming approach for SDN-enabled 5G wireless networks. Netw. IEEE **29**(1), 49–55 (2015)
42. P. Ameigeiras, J.J. Ramos-Munoz, L. Schumacher, J. Prados-Garzon, J. Navarro-Ortiz, J.M. Lopez-Soler, Link-level access cloud architecture design based on SDN for 5G networks. Netw. IEEE **29**(2), 24–31 (2015)
43. S. Talwar, D. Choudhury, K. Dimou, E. Aryafar, B. Bangerter, K. Stewart, in *Enabling Technologies and Architectures for 5G Wireless*. IEEE MTT-S International Microwave Symposium (IMS), pp. 1–4, Jun 2014
44. Y. Mehmood, W. Afzal, F. Ahmad, U. Younas, I. Rashid, I. Mehmood, in *Large Scaled Multi-User MIMO System So Called Massive MIMO Systems for Future Wireless Communication Networks*. Proceedings of IEEE 11th International Conference on Automation and Computing (ICAC), pp. 1–4, Sept 2013
45. E. Larsson, O. Edfors, F. Tufvesson, T. Marzetta, Massive MIMO for next generation wireless systems. IEEE Commun. Mag. **52**(2), 186–195 (2014)
46. X. Chen, R.Q. Hu, Y. Qian, in *Distributed Resource and Power Allocation for Device-To-Device Communications Underlaying Cellular Network*. Global Communications Conference (GLOBECOM), pp. 4947–4952, Dec 2014
47. Z. Zhang, R.Q. Hu, Y. Qian, A. Papathanassiou, G. Wu, in *D2D Communication Underlay Uplink Cellular Network With Fractional Frequency Reuse*. Design of Reliable Communication Networks (DRCN), pp. 247–250, Mar 2015
48. T. Djerafi, O. Kramer, N. Ghassemi, A.B. Guntupalli, B. Youzkatli-El-Khatib, K. Wu, in *Innovative Multilayered Millimetre-Wave Antennas for Multi-Dimensional Scanning and Very Small Footprint Applications*. Proceedings of 6th European Conference on Antennas and Propagation (EUCAP), pp. 2583–2587, Mar 2012
49. J.J. Vegas Olmos, I. Tafur Monroy, in *Millimeter-Wave Wireless Links for 5G Mobile Networks*. Proceedings of 17th International Conference on Transparent Optical Networks (ICTON), pp. 1–1, July 2015
50. S. Scott-Hayward, E. Garcia-Palacios, Multimedia resource allocation in mmwave 5G networks. IEEE Commun. Mag. **53**(1), 240–247 (2015)
51. S. Salsano, N. Blefari-Melazzi, A. Detti, G. Morabito, L. Veltri, Information centric networking over SDN and OpenFlow: architectural aspects and experiments on the OFELIA testbed. Comput. Netw. **57**(16), 3207–3221 (2013)
52. A. Shalimov, D. Zuikov, D. Zimarina, V. Pashkov, in *Advanced Study of SDN/OpenFlow Controllers*. Proceedings of the 9th Central and Eastern European Software Engineering Conference in Russia, Oct 2013
53. N. Cvijetic, A. Tanaka, P. Ji, K. Sethuraman, S. Murakami, SDN and OpenFlow for dynamic flex-grid optical access and aggregation networks. Lightwave Technol. J. **32**(4), 864–870 (2014)
54. Open networking foundation. Interoperability Event Technical Paper, vol. 4, 7 Feb 2013
55. Software-defined networking: a perspective from within a service provider environment. Internet Engineering Task Force, Mar 2014
56. L.L. Wei, R.Q. Hu, T. He, Y. Qian, in *Device-to-Device (D2D) Communications Underlaying MU-MIMO Cellular Networks*. Proceedings of IEEE GlOBECOM, pp. 4902–4907, Dec 2013
57. A. Dixit, F. Hao, S. Mukherjee, T.V. Lakshman, Towards an elastic distributed SDN controller. ACM SIGCOMM Comput. Commun. Rev. **43**(4), 7–12 (2013)
58. P. Bosshart, G. Gibb, H.S. Kim, G. Varghese, N. McKeown, Forwarding metamorphosis: fast programmable match-action processing in hardware for SDN. ACM SIGCOMM Comput. Commun. Rev. **43**(4), 99–110 (2013)

59. M.K. Shin, K.H. Nam, H.J. Kim, in *Software-Defined Networking (SDN): A Reference Architecture and Open APIs*. ICT Convergence (ICTC), pp. 360–361, Oct 2012
60. S. Schmid, J. Suomela, in *Exploiting Locality in Distributed SDN Control*. Proceedings of the Second ACM SIGCOMM Workshop, pp. 121–126, Aug 2013
61. S. Scott-Hayward, G. O'Callaghan, S. Sezer, in *SDN Security: A Survey*. Future Networks and Services, pp. 1–7, Nov 2013
62. White Paper on "Network Functions Virtualisation". http://portal.etsi.org/NFV/NFVWhitePaper.pdf, 13 Oct 2014
63. Software-Defined Networking: The New Norm for Networks. https://www.opennetworking.org/images/stories/downloads/sdnresources/white-papers/wp-sdn-newnorm.pdf, 13 Oct 2014
64. J. Batalle, J. Ferrer Riera, E. Escalona, J.A. Garcia-Espin, in *On the Implementation of NFV Over an OpenFlow Infrastructure: Routing Function Virtualization*. IEEE SDN4FNS, pp. 1–6, Nov 2013
65. H. Masutani, NTT Network Innovation Labs. in *Yokosuka, Japan. Requirements and Design of Flexible NFV Network Infrastructure Node Leveraging SDN/OpenFlow*. IEEE ONDM, pp. 258–263, May 2014
66. M. Palkovic, P. Raghavan, M. Li, A. Dejonghe, A. Dejonghe, L. Van der Perre, F. Catthoor, Future software-defined radio platforms and mapping flows. IEEE Sig. Process Mag. **27**, 22–33 (2010)
67. M. Sadiku, C. Akujuobi, Software-defined radio: a brief overview. IEEE Potentials **23**(4), 14–15 (2004)
68. B.A.A. Nunes, M. Mendonca, X.-N. Nguyen, K. Obraczka, T. Turletti, A survey of software-defined networking: past, present, and future of programmable networks. IEEE Commun. Surv. Tutorials **16**(3), 1617–1634 (2014)
69. S. Sezer, S. Scott-Hayward, P.K. Chouhan, B. Fraser, D. Lake, J. Finnegan, N. Viljoen, M. Miller, N. Rao, Are we ready for SDN? Implementation challenges for software-defined networks. IEEE Commun. Mag. **51**(7), 36–43 (2013)

Chapter 3
Massive MIMO Coordination in 5G Heterogeneous Networks

3.1 Massive MIMO Technology

The Next Generation Mobile Networks Alliance feels that 5G should be rolled out by 2020 to meet the continuously increasing demand for higher data rates, higher energy efficiency, larger network capacity and higher mobility required by new wireless applications. The industry generally believe that compared to 4G, the 5G network should increase the system capacity by 1000 times, enhance the spectral efficiency, energy efficiency and data rates by 10 times (i.e., peak data rate of 10 Gbps for low mobility and peak data rate of 1 Gbps for high mobility), and increase the average cell throughput by 25 times [1]. Potential 5G cellular architectures and key technologies are being investigated, including ultra-dense heterogeneous network, self-organization network, massive MIMO, filter-bank based on multicarrier (FBMC), full duplex technology and so on [2, 3].

MIMO is a method for multiplying the capacity of a radio link using multiple transmit and receive antennas to exploit multipath propagation. As one of the most effective means to improve the system spectrum efficiency, transmission reliability and data rate, MIMO technology has become an essential element of wireless communication standards including Wi-Fi (IEEE 802.11n/ac/ad/aj), WiMAX (IEEE 802.16e), 4G (LTE/LTE-A) and other wireless communication systems. As to the MIMO implementations, the most modern standard, LTE-Advanced, allows for up to 8 antenna ports at the base station and the corresponding improvement is relatively modest. In 2010, Thomas L. Marzetta studied multi-user MIMO (MU-MIMO) systems with multi-cellular time-division duplex (TDD) scenario. In his study, each base station is supposed to deploy a large number of antennas and the massive MIMO concept is given to make a difference from the conventional MIMO system with limited number of antennas [4]. In massive MIMO systems, the BSs are equipped with a very large number of service antennas (e.g., hundreds or thousands) and the total system capacity will approximately linearly increase with the minimum number of transmitter and receiver antennas. With more antenna

© The Author(s) 2016
B. Rong et al., *5G Heterogeneous Networks*,
SpringerBriefs in Electrical and Computer Engineering,
DOI 10.1007/978-3-319-39372-8_3

elements, it becomes possible to steer the transmission towards the intended receiver, which may greatly reduce the interference. Furthermore, the massive MIMO system can acquire the great array gain to reduce the constraints on linearity and accuracy of transmit power and improve the RF power efficiency, which makes massive MIMO system possible to be built with cheap and low power components [5–7].

In a massive MIMO system, different data streams will go through different paths reaching to the receiver at the other end of the air, even the same data stream will behave differently because of different experience in the frequency domain. That means the user data streams radiated from MIMO transmitters are usually shaped in the air by the so-called channel response between each transmitter and receiver at particular frequency. This inherent wireless transmission phenomenon is equivalent to apply a long tap finite impulse response (FIR) filter with unwanted non-flat frequency response on each data stream. And it will introduce frequency distortion and cause poor system performance. To tackle this frequency dependent channel degradation, we can add precoding at the fronthaul/backhaul massive MIMO transmitter side with pseudo-inversed channel response to pre-compensate the channel effects, and then the cascade system will provide 'corrected' spatial multiplexed data streams at different receivers respectively.

3.2 MmWave Hybrid MIMO Precoding

In 5G, millimeter-wave is a promising technology that will bring high-capacity wireless access [8]. A general deployment of mmWave system in 5G mobile networks is shown in Fig. 3.1, where the mmWave small cells cooperate with macro BS to provide seamless coverage and high data rate in most coverage areas. MmWave has the frequency range from 26.5 to 300 GHz, thus the network equipped with mmWave system can enjoy super wide bandwidth. Moreover, mmWave communication system can aim the target user more precisely as it has a much narrower antenna beam size compared to the microwave.

As the antenna number grows, simple linear precoding schemes such as zero forcing (ZF) are virtually optimal and asymptotically approach the sum capacity achieved by dirty paper coding (DPC). In mmWave systems, MIMO precoding is generally different than that at lower frequencies. In conventional precoding schemes, the complex symbols are first modified in both amplitudes and phases at the baseband, and then the processed signals are upconverted to around the carrier frequency after passing through digital-to-analog (D/A) converters, mixers, and power amplifiers (i.e. RF chains). In such schemes, each antenna element needs to be supported by a dedicated RF chain. Practically, it is too expensive to be implemented in mmWave systems due to the large number of antennas.

To address this issue, researchers proposed a hybrid precoding structure with active antennas, which generally controls the signal phase on each antenna via a network of analog phase shifters. In the hybrid precoding structure, as illustrated in

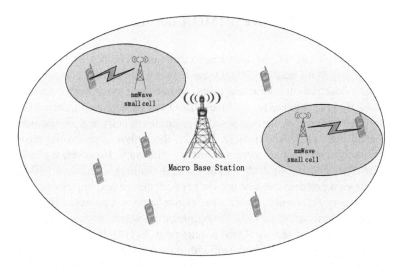

Fig. 3.1 mmWave 5G mobile network architecture

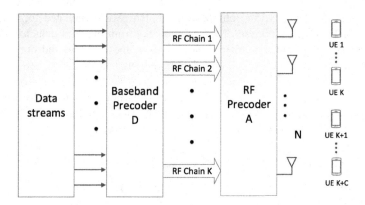

Fig. 3.2 Hybrid mmWave precoding structure

Fig. 3.2, the process is divided into analog and digital domains. User data streams are firstly precoded by the baseband digital processing **D** and transferred by the corresponding RF chains. Then they are mapped into each antenna element by the analog phase processing **A** (RF processing). In this process, the number of RF chains is lower than that of antennas, which reduces the complexity and power consumption in mmWave system. It should be noted that baseband digital processing **D** can make both amplitude and phase modifications, while analog phase processing **A** only modifies the phases with variable phase shifters and combiners.

3.3 The Need of Massive MIMO Coordination in 5G

Massive MIMO is one of the important candidate technologies for 5G mobile networks [9, 10]. With massive MIMO, the network can deploy a large number of antennas at a base station and access point, and achieve much higher performance than that of 4G [11]. Moreover, massive MIMO can be deployed in heterogeneous networks (HetNets) which are composed of a variety of cells, e.g., a macrocell and several small cells [12]. Additionally, in [12], the authors stressed the promising issues on interference management and energy efficiency. However, the cell association for users and resource allocation for base stations in massive MIMO networks becomes a problem due to complex network design and implementation. For example, an energy efficient resource allocation scheme was proposed for a network with large number of antennas [13]. An optimization was formulated and solved to meet per user quality of service (QoS) requirement. In [14], the authors proposed a power control algorithm for a network with massive MIMO and non-cooperative beamforming, which aims to meet QoS requirement of users especially at cell edges.

In 5G heterogeneous networks, cell association is an important issue because users have to associate with the most suitable cell so that their performance is maximized [15]. To analyze the network selection, evolutionary game can be applied [16]. Specifically, an optimization problem was formulated for load balancing which assigns a portion of resource blocks from different cells to different users to meet some fairness criteria. Nevertheless, these studies did not consider independent and competitive cell association and antenna allocation performed by the users and access points, respectively. Notably, if the network lacks centralized coordination and control (e.g., in a distributed environment), the cell association and resource allocation will become more complex and the users as well as the base stations are self-interested to maximize their own benefits.

3.4 Massive MIMO Coordination with SDN

We propose an integration scheme which can implement effective RRM for mmWave system in 5G network. In this scheme, we use null-space based hybrid precoding to overcome the mmWave constrains as well as mitigating inter-tier interference and distinguish our work by involving an SDN controller to handle massive MIMO coordination for the whole 5G network.

As shown in Fig. 3.2, here we assume a downlink mmWave system with K single antenna users. The BS and LPNs are all equipped with massive MIMO containing N_{BS} and N transmission antennas respectively. Define K_j as the number of users of the jth LPN and C_j the number of users interfered by the jth LPN when served by other nodes. Notably, K_j and C_j should meet the constraint

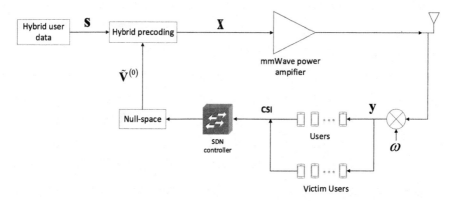

Fig. 3.3 Block diagram of our proposed scheme

$K_j + C_j \leq N$. The stationary channel $\mathbf{H}_j \in \mathbf{C}^{(Kj+Cj) \times N}$ between a certain LPN j and the users it serves is

$$H_j = \begin{bmatrix} H_{j,1} & \cdots & H_{j,K_j} & H_{j,K_j+1} & \cdots & H_{j,K_j+C_j} \end{bmatrix}^H \tag{3.1}$$

where the kth row vector $\mathbf{H}_{j,k} \in \mathbf{C}^{1 \times N}$ is the channel between the jth LPN and the kth user. Specially, $\mathbf{H}_0$ denotes the stationary channel between macro BS and the users it serves. $\mathbf{H}_{0,k}$ denotes the channel between the macro BS and the kth user.

As shown in Fig. 3.3, our processing diagram emphasizes the acquisition of CSI for LPNs and the generation of null-space based hybrid precoding matrix. Here the SDN controller is based on the proposed SDN architecture aforementioned in this book.

Practically, LPNs can only collect the local CSI through the backward channel of their own served users. Therefore, LPNs don't know the CSI of the external victim users interfered by them. In our proposed scheme, however, the hub-spoke structure of the SDN network enables the CSI of all mmWave channels to be collected and disseminated through the backhaul link.

As illustrated in Fig. 3.4, our proposed CSI acquisition method makes BS collect and report the CSI information to the SDN controller. More exactly, BS will send SDN the channel matrix on victim users of each LPN (i.e. sending SDN every $\begin{bmatrix} \mathbf{H}_{j,Kj+1}^H, \cdots, \mathbf{H}_{j,Kj+Cj}^H \end{bmatrix}^H$ corresponding to the jth LPN). Then the SDN controller will perform the block diagonal (BD) algorithm and generate null-space vector $\widetilde{\mathbf{V}}_{j,v}^{(0)}$ for LPN j. Then SDN controller will compute on the null-space vector $\widetilde{\mathbf{V}}_{j,v}^{(0)}$ to achieve precoding matrix by the hybrid precoding algorithm as proposed later. Finally, SDN controller sends the precoding matrix to each LPN through downlink information. In this scheme, the SDN controller takes over all the matrix decomposition computation and precoding. It will be more helpful for the networks where LPNs are too simple to handle the computational work.

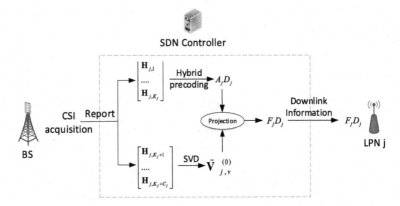

Fig. 3.4 Proposed CSI acquirement methods

Evidently, the interferences from the intended users cannot affect the victim users when the precoding vectors span a subspace of the null-space of the channel vectors from victim users. Accordingly, the hybrid precoding matrix can be constructed by processing the mmWave channel matrix.

The process of hybrid precoding is divided among baseband and RF processing, denoted by $\mathbf{D} \in C^{K \times K}$ and $\mathbf{A} \in C^{N \times K}$ respectively. Each entry of $\mathbf{A}$ is normalized to satisfy $|\mathbf{A}_{a,b}| = \frac{1}{\sqrt{N}}$ where $|\mathbf{A}_{a,b}|$ denotes the magnitude of the (a, b)th element of $\mathbf{A}$. Specifically, $\mathbf{A}_{a,b}$ is the (a, b)th element of $\mathbf{A}$. We perform the RF precoding according to Eq. (3.2), where $\varphi_{a,b}$ is the phase of the (a, b)th element of the conjugate transpose of the stationary channel, i.e. $\mathbf{H}^H$. At the baseband, we observe an equivalent channel $\mathbf{H}_{eq} = \mathbf{HA}$ of a low dimension $K \times K$ where $\mathbf{H}$ is the stationary channel. We use a ZF precoding as the digital precoding algorithm, thus baseband precoding is performed as Eq. (3.3), where Λ is a diagonal matrix introducing for column power normalization. The received signal of the kth user can be calculated by Eq. (3.4), where $\mathbf{s}_j \in C^{Kj \times 1}$ is the transmitted signal vector for a total of K_j users. And $\mathbf{s}_0$ is the transmitted signal vector at BS. $n_{j,k}$ is the additive white Gaussian noise of zero mean and variance σ^2.

$$A_{a,b} = \frac{1}{\sqrt{N}} e^{b\varphi_{a,b}} \tag{3.2}$$

$$D = H_{eq}^H (H_{eq} H_{eq}^H)^{-1} \Lambda \tag{3.3}$$

$$y_{j,k} = \mathbf{H}_{j,k} \mathbf{A}_j \mathbf{D}_j \mathbf{s}_{j,k} + \mathbf{H}_{0,k} \mathbf{A}_0 \mathbf{D}_0 s_0 + n_{j,k} \tag{3.4}$$

For the jth LPN, the complementary space concerning the victim users is given as Eq. (3.5), where $\tilde{\mathbf{H}}_{j,v}$ contains the channel matrix of all the victim users interfered by the jth LPN. To avoid the interference to the victim users, the hybrid precoding matrix should satisfy the condition expressed by Eq. (3.6). And $\widetilde{\mathbf{H}}_{j,v}$ can be further written as Eq. (3.7) by performing SVD.

$$\widetilde{\mathbf{H}}_{j,v} = \left[\mathbf{H}_{j,K_j+1}^H \quad \cdots \quad \mathbf{H}_{j,K_j+C_j}^H \right]^H \tag{3.5}$$

$$\widetilde{\mathbf{H}}_{j,v}\mathbf{A}_j\mathbf{D}_j = \mathbf{0}^{C_j \times K_j} \tag{3.6}$$

$$\widetilde{\mathbf{H}}_{j,v} = \widetilde{\mathbf{U}}_{j,v}\widetilde{\mathbf{\Lambda}}_{j,v}\widetilde{\mathbf{V}}_{j,v} = \widetilde{\mathbf{U}}_{j,v}\left[\sum_{j,v} 0 \right]_{C_j \times N} \left[\widetilde{\mathbf{v}}_{j,v,1}\widetilde{\mathbf{v}}_{j,v,2}\cdots\widetilde{\mathbf{v}}_{j,v,N} \right]^H \tag{3.7}$$

Define the null-space vector $\widetilde{\mathbf{V}}_{j,v}^{(0)} = \left[\widetilde{\mathbf{v}}_{j,v,C_j+1}\cdots\widetilde{\mathbf{v}}_{j,v,N} \right]$. Since the column vectors belonging to $\widetilde{\mathbf{V}}_{j,v}^{(0)}$ locate in the null-space of all victim users, we will have $\widetilde{\mathbf{H}}_{j,v}\widetilde{\mathbf{V}}_{j,v}^{(0)} = 0$. For the jth LPN, define the projection matrix $\mathbf{M}_j$ based on the null-space of victim users by Eq. (3.8). The new analog precoding matrix after projection can be calculated as Eq. (3.9).

$$\mathbf{M}_j = \widetilde{\mathbf{V}}_{j,v}^{(0)}(\widetilde{\mathbf{V}}_{j,v}^{(0)})^H \tag{3.8}$$

$$\mathbf{F}_j = \mathbf{M}_j\mathbf{A}_j \tag{3.9}$$

Define $\mathbf{P}_j$ as the transmit power at the jth LPN satisfying $E[\mathbf{s}_j\mathbf{s}_j^H] = \frac{P_j}{K_j}\mathbf{I}_{K_j}$ where $\mathbf{s}_j$ denotes the signal vector for K_j users in total. We further normalize $\mathbf{F}_j$ to satisfy $\|\mathbf{F}_j\mathbf{D}_j\|_F^2 = K_j$ for the total transmission power constraint. Thus the SINR of the kth user served by LPN j is given as Eq. (3.10), where $d_{j,r}$ denotes the rth column of $\mathbf{D}_j$. To support each user's QoS requirement, the SINR should meet the constraints expressed by Eq. (3.11), $R_{j,k}$ is the data rate demanded by user k and B is the total bandwidth. Then the minimum $\mathbf{F}_{j,k}$ can be solved by linear Eq. (3.12).

$$SINR_{j,k} = \frac{\frac{P_j}{K_j}\left|\mathbf{H}_{j,k}\mathbf{F}_j d_{j,k}\right|^2}{\sigma_{j,k}^2 + \sum_{\substack{r=1 \\ r \neq k}}^{K_j} \frac{P_j}{K_j}\left|\mathbf{H}_{j,k}\mathbf{F}_j d_{j,r}\right|^2} \tag{3.10}$$

$$B\log(1 + SINR_{j,k}) \geq R_{j,k}, \quad j = 1, 2, \ldots J; \ k = 1, 2, \ldots K_j \tag{3.11}$$

$$B\log(1 + SINR_{j,k}) = R_{j,k} \tag{3.12}$$

This null-space based hybrid precoding scheme has the following advantages compared to conventional linear precoding ones.

Reduced hardware and computational complexity: Different from the conventional linear precoding schemes that typically require N RF chains, hybrid precoding reduces the number of RF chains in mmWave system from antenna scale to user scale, which brings a significant decrease on hardware complexity. Thus our proposed hybrid precoding scheme owns a much lower complexity than linear precoding schemes. Practically, our proposed algorithm can significantly reduce complexity and power consumption in 5G mmWave Hetnets.

Coordinated CSI acquisition mechanism aided by SDN: In conventional HetNets communication system, the served user equipment reports its CSI only to the access point it connects. Since it is not responsible for reporting to the other nodes that interferes it, LPNs cannot detect the victim users' CSI. However, by performing our

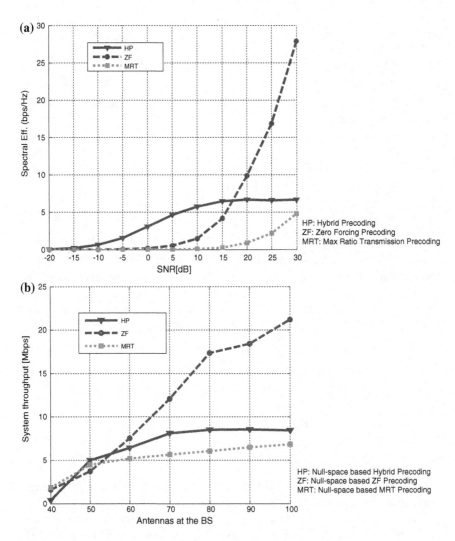

Fig. 3.5 Performance comparison of different MIMO coordination schemes

proposed algorithm, the SDN controller can get enough information form the precoding vector to avoid the interference to the victim users. In addition, by letting SDN perform the null-space construction, the computational burden on LPN can be relaxed.

Reduced interference and improved QoS: By performing hybrid precoding based on null-space of the victim users, the channel matrix of victim users and the generated precoding vectors will be orthogonal to each other, thus the interference to the victim users can be totally avoided. Meanwhile, to further conform to the practice, the SINR calculation is adjusted in order to effectively mitigate the interference and achieve improved QoS for users with satisfying power consumption, cooperating with hybrid precoding algorithm.

We have developed Matlab simulation for the 5G HetNet shown in Fig. 3.1. Figure 3.5a illustrates the spectral efficiency of conventional linear precoding schemes and the proposed hybrid one under different SNR conditions. Apparently the hybrid precoding scheme has a higher spectral efficiency than the MRT precoding, and thus is more suitable for realistic mmWave system. We also simulated our proposed hybrid precoding scheme as well as the linear ZF and MRT by adopting the null-space method to mitigate inner-tier interference. As shown in Fig. 3.5b, we compare these three schemes in terms of system throughput while changing the number of massive MIMO antennas at access points. Massive MIMO demonstrates a performance improvement in system throughput, as the increase of antenna number can eliminate the interference to the victim users. In addition, the null-space based hybrid precoding achieves considerable throughput gain compared to the MRT precoding when antenna number increases. Figure 3.5 overall justifies that our proposed hybrid precoding scheme is able to achieve competitive performance at lower complexity.

References

1. C.-X. Wang, F. Haider, X. Gao et al., Cellular architecture and key technologies for 5G wireless communication networks. IEEE Commun. Mag. **52**(2), 122–130 (2014)
2. X. You, Z. Pan, X. Gao et al., The 5G mobile communication: the development trends and its emerging key techniques. Sci. China-Inf. Sci. **44**(5), 551–563 (2014)
3. S. Chen, J. Zhao, The requirements, challenges, and technologies for 5G of terrestrial mobile telecommunication. IEEE Commun. Mag. **52**(5), 36–43 (2014)
4. T.L. Marzetta, Noncooperative cellular wireless with unlimited numbers of base station antennas. IEEE Trans. Wireless Commun. **9**(11), 3590–3600 (2010)
5. L. Lu, G.Y. Li, A.L. Swindlehurst et al., An overview of massive MIMO: benefits and challenges. IEEE J.-STSP **8**(5), 742–758 (2014)
6. E.G. Larsson, O. Edfors, F. Tufvesson, T.L. Marzetta, Massive MIMO for next generation wireless systems. IEEE Commun. Mag. **52**, 186–195 (2014)
7. F. Rusek, D. Persson, B.K. Lau et al., Scaling up MIMO: opportunities and challenges with very large arrays. IEEE Sig. Process. Mag. **30**(1), 40–60 (2013)

8. J.J. Vegas Olmos, I. Tafur Monroy, in *Millimeter-Wave Wireless Links for 5G Mobile Networks*. Proceedings of 17th International Conference on Transparent Optical Networks (ICTON), p. 1, Jul 2015

9. J.G. Andrews, S. Buzzi, W. Choi, S. Hanly, A. Lozano, A.C.K. Soong, J.C. Zhang, What will 5G be? IEEE J. Sel. Areas Commun. (to appear)

10. P. Demestichas, A. Georgakopoulos, D. Karvounas, K. Tsagkaris, V. Stavroulaki, J. Lu, C. Xiong, J. Yao, 5G on the horizon: key challenges for the radio-access network. IEEE Veh. Technol. Mag. **8**(3), 47–53 (2013)

11. E. Larsson, O. Edfors, F. Tufvesson, T. Marzetta, Massive MIMO for next generation wireless systems. IEEE Commun. Mag. **52**(2), 186–195 (2014)

12. L. Lu, G. Li, A. Swindlehurst, A. Ashikhmin, R. Zhang, An overview of massive MIMO: benefits and challenges. IEEE J. Sel. Top. Signal Process. (to appear)

13. D.W.K. Ng, E.S. Lo, R. Schober, Energy-efficient resource allocation in OFDMA systems with large numbers of base station antennas. IEEE Trans. Wireless Commun. **11**(9), 3292–3304 (2012)

14. J. Choi, Massive MIMO with joint power control. IEEE Wirel. Commun. Lett. (to appear)

15. D. Bethanabhotla, O.Y. Bursalioglu, H.C. Papadopoulos, G. Caire, in *User Association and Load Balancing for Cellular Massive MIMO*. Proceedings of Information Theory and Applications Workshop (ITA), pp. 1–10, Feb 2014

16. D. Niyato, E. Hossain, Z. Han, Dynamics of multiple-seller and multiple-buyer spectrum trading in cognitive radio networks: a game theoretic modeling approach. IEEE Trans. Mob. Comput. **8**(8), 1009–1022 (2009)

Chapter 4
Conclusion and Future Research Directions

4.1 Concluding Remarks

5G has to be more massive and flexible to enable large scale of services and scenarios. Despite the advances achieved by 4G mobile network, novel market trends are imposing unprecedented challenges on 5G. SON has been widely accepted as one of the most promising technologies to meet the latest demand of mobile data communications. In this book, we present the state-of-the-art of 5G self-organizing and optimization technologies. Chapter 1 introduces the smart LPN as a fundamental technology to achieve high performance with 5G SON. Since physical-layer channel is crucial to performance evaluation, we present a 3D MIMO channel model and use it to analyze the smart LPN in 5G HetNets. Chapter 2 proposes a SDN/NFV based SON architecture, where OpenFlow protocol is employed to enhance the network flexibility by separating packet forwarding (data path) and routing decisions (control path). Further with NFV, the network functions are decoupled from their corresponding physical equipment to improve the scalability of network deployment and achieve the ultimate goal of SON. Chapter 3 presents the latest research result of massive MIMO coordination for 5G heterogeneous networks. Due to the trend of network densification, MIMO coordination becomes essential to the paradigm of self-organizing and optimization. We, correspondingly, give detailed elaboration on MIMO precoding, interference cancellation and SDN controlled architecture design. We also demonstrate simulation results to justify the performance of the proposed schemes from multiple aspects.

© The Author(s) 2016
B. Rong et al., *5G Heterogeneous Networks*,
SpringerBriefs in Electrical and Computer Engineering,
DOI 10.1007/978-3-319-39372-8_4

4.2 Future Works

The evolving 5G mobile networks are envisioned to provide higher data rates, enhanced end-user QoE, reduced end-to-end latency, and lower energy consumption. With the standardization process of 5G, it is expected that more research will be carried out under the structure of SON by adopting SDN and NFV. Meanwhile, the widely acknowledged physical layer technologies in 5G, such as massive MIMO and millimeter wave communications will also contribute to the realization of SON. We highlight some important topics for future 5G self-organization and optimization as follows.

- *Big data into SON*: 5G SON is challenged to cope with network densification, split of control and data plane, network virtualization, simultaneous use of different medium access control and physical layers, and flexible spectrum allocations. Big data can provide SON with the function of machine learning and data analytics, and further achieve network level proactive intelligence. In future system, the data center collects a large amount of measurement and control information during normal operation of 5G networks. Big data and cloud computing offer a significant opportunity for service providers from both business and network optimization perspectives.
- *Network convergence*: One promising feature of 5G system is the unified agnostic solution of networking, as a result of the convergence of LTE, WiFi, WiMAX, etc. With hybrid access strategy, 5G network will be able to offer increased capacity essential to handle peak traffic of residential users. 5G radio will increasingly complement and overlap with traditional fixed-broadband access, forming a deeply-converged heterogeneous system. As radio access technology evolves, 5G has to face the challenge of adopting new radio frequencies and transmission schemes.
- *3D mmWave MIMO channel for urban areas*: There is a fast-growing interest in investigating mmWave bands from industry and research community. Future study of 5G mmWave channel model may concentrate on the fundamental research into measurement, analysis and characterization of propagation models for urban areas. In particular, 2D arrays will be employed at mmWave to formulate distance-dependent elevation modeling, with special emphasis on using the ray tracer to determine elevation model parameters.

Printed in the United States
By Bookmasters